Oenology in Practice

George Z. Kyzas • Fragkiskos Papageorgiou

Oenology in Practice

Chemistry of Wines and Winemaking Protocols

Springer

George Z. Kyzas
Hephaestus Laboratory
School of Chemistry, Faculty of Sciences
Democritus University of Thrace
Kavala, Greece

Fragkiskos Papageorgiou
Hephaestus Laboratory
School of Chemistry, Faculty of Sciences
Democritus University of Thrace
Kavala, Greece

ISBN 978-3-031-85533-7 ISBN 978-3-031-85531-3 (eBook)
https://doi.org/10.1007/978-3-031-85531-3

This Springer imprint is published by the registered company Springer Nature Switzerland AG
The registered company address is: Gewerbestrasse 11, 6330 Cham, Switzerland

In the chaos of knowledge we draw paths.
Dedicated to my world—Sotiria, Alexandra and Achilleas.
—Fragkiskos Papageorgiou

Dedicated to my sons (Zacharias and Charis) and my wife Vicky.
—Prof. George Z. Kyzas

Preface

This book is the result of several years of study. The academic combined with the professional engagement in the sciences of chemistry and oenology led us to this result. This is a book that can be easily read by either students of oenology or professional winemakers and oenologists. There is a well-known saying that good wine is created in the vineyard. We disagree with this saying that good wine is made in a laboratory. However, the truth lies somewhere in between. To create a good wine, we need at least a high-quality raw material. With our expertise, analyses, and the use of suitable oenological additives and practices, we can produce a good wine. Based on the above, it is clear that our book focuses exclusively on the science of oenology, with little emphasis on the science of viticulture.

Our task is to create a decent wine, even with mediocre or poor-quality raw materials. Climate change, whether we like it or not, has greatly affected the quality of crops, including grapes. Anyone in the field of oenology is well aware of this. Ideal vintage years are becoming rarer, and studies show that in the future things will become more and more difficult in this regard. The planet is changing, and we must adapt to this change. Science evolves every day, and many new tools are being developed to deal with difficulties that were previously difficult or impossible to deal with. The science of oenology is evolving, segmenting, and specializing. As researchers, it is our duty to develop and disseminate knowledge.

All of the above prompted us to create this book. The first chapter analyzes the chemistry of wine, while the second chapter presents winemaking protocols that we have worked with and recommend for all types of wine. In the third chapter, we provide a step-by-step guide to the most basic analyses required during production. The final chapter examines the oenological additives that can be used in the production of wine to implement these protocols or address problems that may arise during production.

In this book, we aim to make the lives of professional winemakers, oenologists, and oenology students a little easier. The science of oenology is so complex that even professionals with years of experience in this field require a reference manual to consult at any time, in order to address the various questions and problems that arise during the production of wine. In the first chapter, a brief overview is made of the history of wine and its evolution into a science is provided. This is followed by a detailed theoretical explanation of the chemistry of wine, serving as a reference point for readers to revisit and clarify any complex chemical compounds mentioned later on. In the third chapter, winemaking protocols are presented for the first time at an academic level, which, if followed faithfully, will yield a very good result. They are protocols that we have worked on with excellent results. Additionally, the same chapter includes a protocol for creating a wine that many have likely always wanted to produce but lacked the knowledge to achieve. In the fourth chapter, detailed protocols for all the chemical analyses necessary at various stages during winemaking are presented. In the fifth chapter, all commercially available materials are presented, which can be used to implement winemaking protocols and correct errors that may arise at various stages. Additionally, materials used to improve and stabilize the final result are also presented. In the final and sixth chapter, the fundamental stages of wine tasting are presented, enabling readers to understand how wine is tasted and evaluate correctly the final result of all the preceding stages. From the above, it becomes clear that the manual you are holding or reading on your computer screen is a complete guide to the science of oenology. It is oenology in practice.

Kavala, Greece George Z. Kyzas
Kavala, Greece Fragkiskos Papageorgiou

Contents

Introduction

1

1.1 Historical Evidence

According to archeologists and paleontologists, the vine, from which the wine comes, has a prehistory of several million years. Scientific proof of the age of the vine is fossilized vine plants that are 60 million years old [1]. Even before the ice age, it thrived in the polar zone, mainly in the region we know today as Iceland, in Northern Europe and also in northwestern Asia. The glaciers had a significant effect in limiting its spread and in some way imposed the geographical isolation of many varieties, some of which evolved into different species. In time, several populations of wild vines moved to warmer zones, mainly to the wider area of the southern Caucasus. In this region, between the Black Sea, the Caspian Sea and Mesopotamia, the species Vitis vinifera was born. Different varieties of this species are still cultivated today.

In the effort to find a time chain that has a scientific basis, we could, based on the archeological discoveries, say that the first remains of wine in a container were found in Henan Province, China and are 9000 years old, while the previous find in Hajji Firuz Tepe, Iran was 7000 years old, and the one immediately preceding it, from the same area, was 5100 years old (Fig. 1.1).

The cultivation of the vine and winemaking is thought to have its roots in the agricultural revolution and the permanent settlement of populations for the purpose of cultivation, i.e., it dates back to around 5000 BC [2]. The ancient Persians, the Semitic peoples, and the Assyrians are considered to be among the first known grape growers [3]. More generally, one could say that the greatest of the cultures of the Eastern Mediterranean have an important role in the development of viticulture and wine production.

The cultivation of the vine for winemaking dates back to somewhere between 4000 and 6000 BC. near the Caspian Sea. The first wine will be discovered by chance, most likely around 10,000–8000 BC, when someone drank the fermented juice of wild grapes that had been collected and stored [4].

G. Z. Kyzas, F. Papageorgiou, *Oenology in Practice*,
https://doi.org/10.1007/978-3-031-85531-3_1

Fig. 1.1 Wine vessel founded in Hajji Firuz Tepe

Legend says that all started in Persia, where King Jasmid loved grapes and stored them in clay pots to last through the winter. At some point—according to the legend—after time passed, the grapes were no longer tasty and the liquid inside the container that preserved them no longer had any sweetness. So they thought that the containers that once held delicious grapes and sweet must turned into a poisonous liquid. A woman from the king's harem, suffering from headaches, decided to end her life and thought of drinking the poison contained in the clay pots. But instead of dying as he had originally decided he fell into a deep sleep and when he woke up he felt much better. This fact was conveyed to the king and from then on he and his entourage decided to drink this new drink [5].

The latest archeological discoveries are bringing to the surface evidence related to wine in various places around the world. In China (7000 BC), Georgia (6000 BC), Iran (5000 BC), Greece (4500 BC) Sicily (4000 BC) [6]. The oldest evidence of wine production has been found in Armenia and dates back to 4100 BC. These findings were discovered in a cave named Areni-1 in Vayots Dzor, Armenia. Inside the cave there is a grape press of the time, containers where the alcoholic fermentation took place, storage containers and glasses. Also in the same place were found vine roots and grapes. Logically, the fact that all these technological findings date back to 4100 BC prove that in the region wine production has been taking place since much earlier, but no one can responsibly say when [7]. The discovery of this particular archeological find led one of the biggest television networks like CNN to quote verbatim the following "Forget France, it seems the real birthplace of wine is probably in a cave in Armenia" (Fig. 1.2) [8].

Fig. 1.2 Inside Areni-1 caves in Vayots Dzor, Armenia

In Mesopotamia and Egypt, although vinification takes place, large quantities of wine are imported from elsewhere, e.g., from Palestine and Syria [9]. It seems that in Syria viticulture started as early as 5000 BC [10], while they have been identified since 3000 BC in Jericho. Climatically, the environment of Syria was suitable for viticulture (cold winters and long hot summers). Already at the beginning of the first millennium BC, with the rise to power of the seafaring Phoenicians, in the city-states of Byblos, Beirut, and Sidon, Lebanese wines were traded extensively in the Mediterranean and elsewhere.

1.1.1 The Historical Relationship of Greek Culture with Wine

The appearance of the Minoan civilization in Crete at the same time as the Egyptian civilization developed yet another source of demand for wine. Here too beer was the most popular drink, but as early as 1700 BC the expansion of wine had begun, while already in the fifteenth century BC, at the time of the Mycenaean conquest of Crete, viticulture was already established on the island [11]. By the fifteenth century BC, with the expansion of Greek Mycenaean power to the east and south, viticulture also expanded to mainland Greece [12]. On Linear B tablets discovered in a wine cellar in a palace in Pylos, there is a reference to the god Dionysus, who was later to become known throughout the Greek world as the god of wine. The cult of Dionysus appeared in Thrace. There are many myths and legends associated with Dionysus, but the main elements in his mythology are that he was the son of Zeus and Semele, daughter of Cadmus, the king of Thebes [13]. He was raised by nymphs on the slopes of the mythical mountain of Nysa, where he was trained by the muses and discovered how to make wine from grapes. Worth mentioning is the fact that in one of the countless Greek traditions. There is a Greek tradition that mentions that Dionysos left Mesopotamia because its inhabitants preferred beer (Fig. 1.3).

Fig. 1.3 Sculpture of Greek god Dionysus. The ancient god of wine

The emergence of the city-states (Athens, Argos, Corinth, Sparta) marked the beginning of Greece as a major power in the ancient world. By that time a wide variety of wines were produced throughout the country [14]. Each region, each island had its own style and reputation. For example, the wines of Chios, Lesbos, or Samos were famous as were the black wines from Kos, which were sometimes sold mixed with sea water. The wines from Rhodes were also known to be strong, although less good than those from Lesvos. The philosopher Democritus, who proudly said that he knew all the varieties of the vine, was the only one who believed that these varieties could be counted (Fig. 1.4).

All other authors claim that the list is vast, never ending. This truth becomes more apparent when we consider the different types of wine. It is not possible to mention them all but only the most famous of them. There are as many types of wine as there are regions that produce wine, says Pliny in Natural History 14.4. He himself mentions in natural history 14.8 that it is frivolous to concern ourselves with the enumeration of varieties since the same vine in different places differs. In other words, the perception of his time was so developed that they had observed the first samples of today's so-called terroir [14].

In the middle of the fifth century BC, the first wine laws were passed on the island of Thassos, which regulated the minimum quality of wines that could be sold. Later laws (425–400 BC) deal mainly with the date for harvesting grapes thus ensuring that only ripe grapes are used for winemaking. One could say that the first wine

Fig. 1.4 Bust of the philosopher Democritus

legislation is making its appearance. Of course, we should know that in 1700 BC King Hammurabi. It determined the selling price as well as the period when the wine was allowed to be consumed, limiting it to the harvest season. Another example of legislation comes during the New Dynasty (Egypt 1580–1085 BC), the cultivation of the vine has become so important that the amphorae often have an inscription stating the origin of the wine, the name of the vintner and the name of the pharaoh who reigned, that is, they determined the year of production, just as is done in today's labels with the year of production [15].

We know from many descriptions in Greek and Roman literature that Greek wines were sweet and strong, i.e., with a high-alcohol content. This was one of the reasons why the wine of Ancient Greece was always drunk diluted with water in a ratio usually 1:3 (one part wine to three parts water). The word "wine" denotes precisely the wine mixed with water, while the term "άκρατος – akratos" meant the unadulterated wine. They had special utensils both for mixing (craters) and for cooling it. Drinking wine that had not been mixed with water was considered barbaric and was only practiced by the sick or while traveling as a tonic. The consumption of wine with honey and the use of various herbs were still widespread. Adding absinthe to wine was also a known method (Hippocratic Wine) as was the addition of resin.

The ancient Greek landscape was ideal for the cultivation of the vine since Greece belongs to the zone-habitat of the wine-bearing Eurasian vine or more simply the vine that created the wine culture of European history (Vitis Vinifera). A pioneer in this art, as in many others, Greek antiquity bet on the maximum and highest quality possible production of the vineyard. Thanks to its long tradition, the Greek vintner of the time knew that the future of the vineyard was judged first of all by prediction rather than intervention, by his skill and dedication to his cultivated land. If he did not take care of his vines, there was a good chance that they would become unfruitful and even perish forever. The viticulturist knew that he had to study all the environmental factors and foresee all the possible consequences that would arise from the combination of his choices.

The Greeks transmitted their knowledge and experience of viticulture and wine everywhere they colonized. The Greeks arrived in Sicily in the eighth BC century. Sicilian wines (such as Mamertime) were famous. From Sicily, the Greeks moved to southern Italy (where Etruscan wines already existed. On the coast of France, the Greeks founded the city of Marseille around 600 BC. But it is possible that local varieties had already been cultivated there since the Celts.)

1.1.2 The Roman Empire and Its Role in Relation to Wine

During the Roman Empire, the Greek influence on the creation of wines remained significant. The wines were heavy and sweet, the best of them aged for decades in amphorae. Many Roman writers praised Greek wines from Lesbos, Chios, and other islands for centuries [16]. Trade in Greek wines continued throughout the Roman invasion. The Romans also contributed to the development of the art and later the science of oenology, since they invented the wine press. The grapes were collected on a stone slab and a heavy stone was rolled over them and crushed. The juice was collected from openings in the bottom of the bucket. The Romans also invented the Dolium, a very large clay jar (300 gallons). It was used for aging and storage (usually buried up to its neck) as well as for transporting wine [17] (Fig. 1.5).

Although most Roman wines were white, residual sugars and age gave them an orange color. Also the dark yellow color was sometimes made by smoking the wine.

Fig. 1.5 Dolium, a very large clay jar (300 gallons)

Wine production rose to prominence in Europe mainly with the expansion of the Roman Empire across the Mediterranean, when many important wine-producing regions developed, which still exist today. Even then, winemaking was a meticulous agricultural economy that promoted the development of different varieties and cultivation techniques. Barrels for storage and transport made their appearance, bottles were used for the first time and a rudimentary appellation of origin system developed as certain regions gained a reputation for their fine wines. As winemaking was gradually upgraded, its popularity grew and wine shops became a common feature in many cities throughout the then Roman Empire.

1.1.3 Europe and Wine

Over the centuries, the art of winemaking spread to many European countries such as France, Spain, Germany, and parts of Great Britain. So wine is considered an important part of the daily diet and Europeans began to prefer stronger, heavier wines. Europe's respect for wine was maintained during the Middle Ages. One of the main reasons in part was the dubious quality of the drinking water, so wine was the preferred alternative to accompany one's meal. At the same time, viticulture and winemaking were developing and monasteries throughout the European continent played an important role, which gave impetus to the creation of some of the best vineyards in Europe. The Benedictine monks, for example, became some of the largest wine producers with vineyards in the Campania, Burgundy, and Bordeaux regions of France as well as the Rheingau and Franconia regions of Germany [16] (Fig. 1.6).

Traders and the noble classes consumed wine at every meal and kept well-stocked cellars. During the sixteenth century, wine became a more sophisticated alternative to beer, and as wine products began to diversify, consumers began to find the idea of variety in their drinking habits attractive. They began to discuss the virtues and passions associated with wine with greater fervor than in previous centuries. For example, Shakespeare pointed out that "good wine is a good friend, when well treated," implicitly commenting on the abuse of wine in his day. During this time, London experienced access to clean drinking water, a major development that ushered the wine industry into a new era.

New improved winemaking techniques in the seventeenth and eighteenth centuries resulted in better quality wines, the use of glass bottles with corks and the invention of the corkscrew. In other words, it seems that an industry is being prepared around wine and as is usually the case in such cases there is rapid growth. As it did, the French wine industry took off, with the red wines of the Bordeaux region receiving great acclaim from merchants in the Netherlands, Germany and Scandinavia. Trading wines from Bordeaux against coffee and other sought-after items from the New World helped establish wine's role in the emerging world trade (Fig. 1.7).

While one would imagine that the following century (the nineteenth century) would be the golden age of wine, unfortunately for many regions this did not become

Fig. 1.6 Artistic paint of Benedictine Monks in their lab

Fig. 1.7 Vineyards in Bordeaux, France

reality as Phylloxira made its appearance. More specifically around 1863 many French vineyards were affected by a disease caused by the phylloxera aphid, which sucked the sap from the roots. The solution to the problem came from the other side of the Atlantic, after it was noticed that vines in America were resistant to phylloxera. Thus it was decided to plant American vines in the affected French regions. This created hybrid grape varieties that produced a better variety of wine (Fig. 1.8).

Over the past 150 years, winemaking has undergone a revolutionary evolution as both an art and a science. Great discoveries, research and innovations, created wines of high esthetics. Having access to refrigeration made it easier for wineries to control the temperature of the fermentation process and produce high-quality wines in hot climates. The introduction of harvesting machinery allowed wine producers to increase the size of their vineyards and make them more productive. While the wine industry faces the challenge of meeting the demands of an ever-expanding market without losing the distinctive character of its wines, technology helps ensure a steady supply of quality wines. Contemporary oenology honors the timeless art of winemaking and demonstrates the importance of wine in the history and diversity of European culture.

Wine in time acquired great commercial value and people became increasingly concerned with its development in order to gain a comparative advantage over its competitor. This competition constantly improved the quality of the wine, new

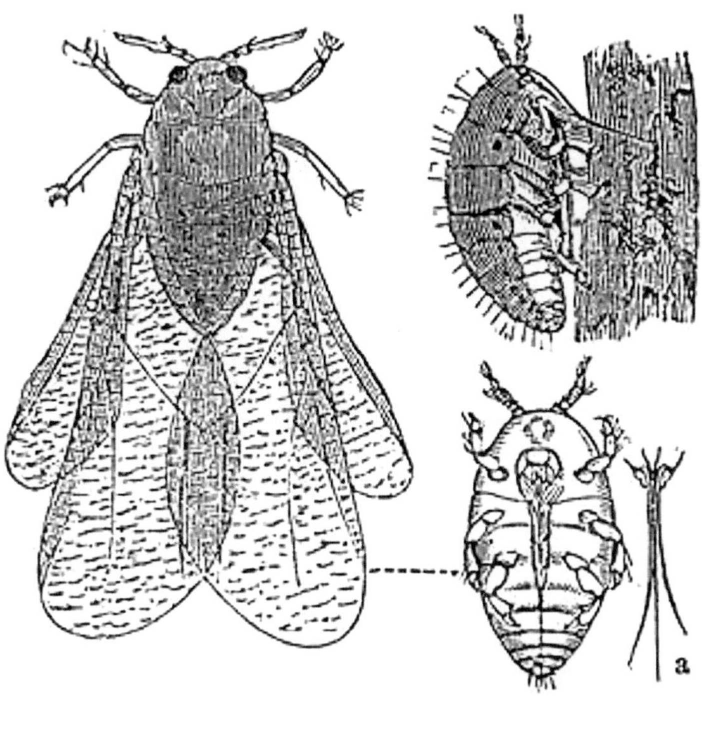

Fig. 1.8 Genus of hemi-pterous insects, typical representative of the Phylloxeridae family

cultivation techniques appeared, new production technologies, innovations that ensured preservation during its transport. The wine revolution had just begun and was constantly evolving. In the midst of all this evolution and continuous improvement, science about 200 years ago made its own revolution and contributed the most to the production of wine [18].

That's enough we are in today's era, the era where we have now decoded every stage of production and have answered many unanswered questions of the past. Technology, the development of sciences, research, and the wine industry are now harmoniously intertwined. Wine has now approached perfection and most technological developments focus on how much easier it is to make an excellent wine rather than how to make an excellent wine. This has been answered for many years now. Knowing all of the above and having dedicated our lives to wine and the science that surrounds it, we created this book with this particular structure. If one knows the basic chemistry of wine (see Chap. 2), the step-by-step operating methodology during production (see Chap. 3), the basic analyzes required (see Chap. 4), and the oenological additives to use (see Chap. 5), then one can to create an excellent wine.

1.2 World Wine Production

According to the latest figures from the OIV, in 2022 [19], despite the heat waves affecting various regions globally, the volume of wine production is anticipated to align with the previous year's level. This marks the fourth successive year of global production being marginally below the average [19]. Specifically based on data from 29 countries, accounting for 91% of global production in 2021, the estimated world wine production for 2022 (excluding juices and musts) ranges from 257.5 to 262.3 million hectoliters, with a median estimate of 259.9 million hectoliters. However, in order to have a more comprehensive picture of the global wine

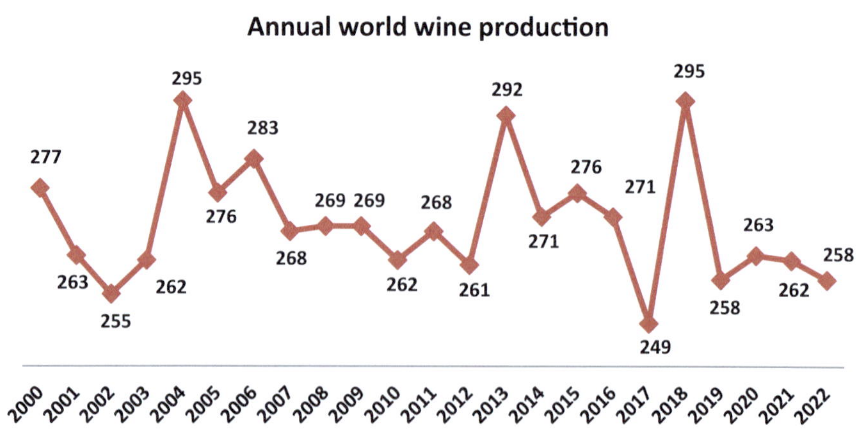

Fig. 1.9 Annual must production in million hectoliters

production, we quote Fig. 1.9 in which the annual must production in million hectoliters is presented from 2000 to 2022. From the numbers we understand the importance and impact that wine has on the global economy.

Based on the information collected on 29 countries, which represent 91% of the world production in 2021, 2022 world wine production (excluding juices and musts) is estimated between 257.5 and 262.3 mhl, with a mid-range estimate at 259.9 mhl. The 2022 wine production volume can be considered slightly below its 20-year average and seems to have fallen by 1% compared to 2021. This is due to higher than-expected harvest volume in Europe (despite the drought and heat waves during spring and summer) and average production level recorded in the Southern Hemisphere and in the USA. Overall, in 2022, the dry and hot conditions observed across different regions of the world have led to early harvests and average volumes; nonetheless an overall good quality is expected. As shown in Fig. 1.9, world wine production is stable around 260 mhl for four consecutive years. It should be noted, however, that the figures for 2022 must be taken with caution as there are still large countries like China and Russia for which information is not available yet. Also, the high volatility in production volumes observed over the last years at regional levels makes the forecasting exercise even more difficult.

In order to achieve a more correct analysis, we will divide Table 1.1 of 29 countries into three zones. The first zone includes the countries of the European Union, the second zone includes major countries in the northern hemisphere outside the European union, and finally the third zone includes major countries in the southern hemisphere.

Throughout the 2022 growing season, the European Union (EU) experienced a series of adverse weather events, including spring frost, hail, excessive heat, and drought. Spring and summer heat waves across Europe have resulted in early ripening. The lack of significant grape diseases and the absence of late summer rains compensated, leading to higher than anticipated yields in various regions and countries. The result is an overall early harvest and an estimated wine production volume of 157 mhl, excluding juices and musts. This volume shows an annual increase of 3.5 mhl (+2%) compared to 2021. Despite the drought affecting some areas, 2022 was a favorable year for wine production in Italy and France, the two largest EU producers, largely due to the late summer rains. Italy maintained its position as the world's leading producer with 50.3 million hectoliters, estimating a volume consistent with its 2021 wine production. France, following the notably low yield of 2021 marked by significant damage from early frosts in April, subsequent summer rains, hailstorms, and mildew, now stands as the EU country with the highest growth rate compared to the previous year. It anticipates a production level of 44.2 million hectoliters, signifying a +17% increase from 2021. The dry and warm conditions this year have lessened disease prevalence in vineyards and resulted in an earlier commencement of grape harvesting. In 2022, Spain is projected to be the world's third-largest wine producer, with an estimated output of 33.0 million hectolitres (mhl). This volume, which is 6% less than in 2021 and 12% below the 5-year average, is largely due to drought conditions and restricted water availability in various regions. Among the other major wine-producing countries in the EU, positive performances

Table 1.1 Wine production of 29 countries in mhl for period 2017–2022

In mhl	2017	2018	2019	2020	2021	2022	Average
Italy	42.5	54.8	47.5	49.1	50.2	50.3	49.1
France	36.4	49.2	42.2	46.7	37.6	44.2	42.7
Spain	32.5	44.9	33.7	40.9	35.0	33.0	36.7
USA	24.5	26.1	25.6	22.8	24.1	23.1	24.4
Australia	13.7	12.7	12.0	10.9	14.8	12.1	12.7
Argentina	11.8	14.5	13.0	10.8	12.5	11.4	12.3
Chile	9.5	12.9	11.9	10.3	13.4	12.4	11.7
South Africa	10.8	9.5	9.7	10.4	10.6	10.2	10.2
China	11.6	9.3	7.8	6.6	5.9	NA	8.2
Germany	7.5	10.3	8.2	8.4	8.7	8.9	8.7
Portugal	6.7	6.1	6.5	6.4	7.3	6.7	6.6
Russia	4.5	4.3	4.6	4.4	4.5	NA	4.5
Romania	4.3	5.1	3.8	3.8	4.5	4.6	4.4
Brazil	2.9	3.0	3.0	3.3	2.7	3.8	3.1
New Zealand	3.6	3.1	2.2	2.3	3.6	3.2	3.0
Hungary	2.9	3.6	2.7	2.9	2.6	2.5	2.9
Austria	2.5	2.8	2.5	2.4	2.5	2.3	2.5
Greece	2.6	2.2	2.4	2.3	2.4	1.7	2.3
Moldova	1.8	1.9	1.5	0.9	1.4	1.3	1.5
Bulgaria	1.2	1.1	0.9	0.8	0.8	0.9	1.0
Georgia	1.0	1.7	1.8	1.8	2.1	2.1	1.8
Switzerland	0.8	1.1	1.0	0.8	0.6	1.0	0.9
Slovenia	0.6	0.9	0.8	0.7	0.6	0.7	0.7
Croatia	0.7	1.0	0.7	0.8	0.5	0.6	0.7
Uruguay	0.8	0.7	0.6	0.7	0.7	0.8	0.7
Czech Rep	0.6	0.7	0.5	0.6	0.6	0.6	0.6
Slovakia	0.3	0.4	0.3	0.4	0.3	0.3	0.3
Luxemburg	0.1	0.1	0.1	0.1	0.1	0.1	0.1
Cyprus	0.1	0.1	0.1	0.1	0.1	0.1	0.1

with respect to 2021 have been recorded in Germany (8.9 mhl, +2%/2021), Romania (4.6 mhl, +4%/2021), Bulgaria (0.9 mhl, +5%/2021), Slovenia (0.7 mhl, +15%/2021), and Croatia (0.6 mhl, +12%/2021). In these countries, the dry and hot growing season has been beneficial for vineyards, a large part of which have been very productive, despite the hot temperatures. Conversely, numerous EU countries are anticipating negative growth compared to 2021. For instance, Portugal, which ranks as the fifth largest wine producer in Europe, is projected to have a wine production volume of 6.7 million hectoliters in 2022, marking an 8% decrease from the previous year. The harvest in Portugal, as in many other European Union countries, has suffered due to the excessive heat during the summer, along with scant rainfall. Nonetheless, it is noteworthy that the anticipated yield for 2022 aligns with the 5-year average. Hungary's estimated wine production is 2.5 mhl in 2022, a level 3% below 2021 and 15% lower than its 5-year average. In this instance, the extreme temperatures noted during spring and summer resulted in an early and comparatively modest harvest. Likewise, Austria, producing 2.3 million hectoliters, forecasts a wine production that is 6% below that of 2021 and 8% beneath its 5-year average. Greece was significantly impacted by this season's heat wave, with its wine production in 2022 anticipated to be just 1.7 million hectolitres (mhl). This not only marks a sharp 29% decrease from the previous year but also a substantial drop from the 5-year average. Meanwhile, the Czech Republic's expected wine production is 0.6 mhl, which is 8% less than in 2021, aligning with its 5-year average.

Second zone includes major countries in the northern hemisphere outside the European union in the top of this zone we have USA. Preliminary estimates for 2022 wine production in the USA, the fourth-largest producer in the world, are 23.1 million hl. This number is 6% less than the 5-year average and 4% less than the previous year. This relative decline can be partly ascribed to early frost damage, summer dry conditions, and the ensuing scarcity of water in some wine regions. Georgia's wine output is predicted to reach a record-breaking level of 2.1 million hectoliters, matching the already substantial production of 2021 (+2%). This is in comparison to other Eastern European countries. This volume is the result of both a government subsidy program that has driven production to record levels and favorable weather that produces high grape harvests. Moldova is expected to produce 1.3 million hectoliters of wine in 2022, a 7% drop from 2021 levels. This year's amount of Swiss wine production is not only higher than last year's, but also 12% more than the average for the previous 5 years. The heat waves that swept through Europe proved advantageous for the comparatively high-altitude Swiss vineyards.

Finally, we have the third zone includes major countries in the southern hemisphere. Preliminary estimates of wine production around this time of year are typically more accurate in the Southern Hemisphere, where harvests conclude in the first trimester of 2022. Following a notable reduction in wine production in 2020 due to unfavorable weather and a record-breaking harvest in 2021, the Southern Hemisphere's projected wine production for 2022 is approximately 55 million hectoliters, a 7% decrease from the year before but fully consistent with its previous 5-year average. In total, it is anticipated that 21% of the world's wine output would come from the Southern Hemisphere in 2022. Every significant wine-producing nation in South America has seen a decrease in production compared to 2021. With a peak wine output of 12.4 mhl in 2022—thanks to dry conditions—Chile is the leading producer in the Southern Hemisphere. This year's production was only 7% lower than the record-breaking high of the previous year, which was 7% higher than its 5-year average. Argentina's wine output fell by around 1.0 million hectoliters in 2022 as a result of highly unpredictable weather (heavy downpours, freezing temperatures, etc.) the production finally reached 11.4 million hectoliters (−9%/2021). Brazil is expected to produce 3.2 million hectoliters of wine in 2022. Despite the dry spring and droughty summer, the projected volume for this year is higher than the average for the previous 5 years, despite the fact that it is 10% less than for 2021. Wine output in South Africa is predicted to reach 10.4 million hectoliters in 2022, a 4% decline from 2021. It should be highlighted, nevertheless, that this number exactly matches its 5-year average. Throughout the Southern Hemisphere, Australia had the biggest decline from the previous year. Australia records a −18% decrease from 2021 with 12.1 mhl. This can be attributed to chilly springtime weather, copious summer rains, and winemakers' seasonal tweaks following the record-breaking 2021 vintage. Once again, in the Southern Hemisphere, New Zealand stands out as an outlier. While New Zealand was the only significant country in the Southern Hemisphere to have a below-average harvest of wine grapes in 2021, it saw record-high wine output in 2022, reaching 3.8 mhl (+44%/2021). The combination of exceptional weather and strong global demand has undoubtedly led to this record-breaking harvest volume [1].

The detailed data used for the analysis can be found in Table 1.1. Specifically, Table 1.1 provides the annual wine production by country from 2017 to 2022. The last column of the table is the 6-year average 2017–2022.

By studying the figures regarding the world wine production, we can come to many conclusions. It is very important that we have a more general picture of wine production since this way we can improve and adapt our winemaking techniques and protocols in difficult years. It is not always necessary for a scientist to invent new tactics to deal with problems that arise during production. Usually our problem has been solved by others. What is enough is to search the literature, to ask, to collaborate, and if we do not find a ready solution, then it is enough to research, experiment and innovate and in this way we will face the new challenges that will come. What we have tried to do in this book after doing all of the above for years, is to offer worked winemaking protocols that ensure excellent results.

References

1. Harutyunyan M, Malfeito-Ferreira M (2022) The rise of wine among ancient civilizations across the Mediterranean basin. Heritage 5:788–812. https://doi.org/10.3390/heritage5020043
2. McGovern PE (2009) Uncorking the past: the quest for wine, beer, and other alcoholic beverages. University of California Press, Berkeley
3. Unwin T (1996) Wine and the vine: an historical geography of viticulture and the wine trade. Routledge, London
4. Begos K (2018) Tasting the past: the science of flavor and the search for the origins of wine. Algonquin Books, Chapel Hill
5. Pellechia T (2006) Wine: the 8,000-year-old story of the wine trade. Running Press, London, pp XI–XII. ISBN: 1-56025-871-3
6. Pellechia T (2006) Wine: the 8,000-year-old story of the wine trade. Thunder's Mouth Press, New York
7. McGovern PE, Fleming SJ, Katz SH (eds) (2003) The origins and ancient history of wine (Food and nutrition in History and Anthropology series no. 11). Routledge, London
8. Hovsepyan R, Willcox G (2008) The earliest finds of cultivated plants in Armenia: evidence from charred remains and crop processing residues in pisé from the Neolithic settlements of Aratashen and Aknashen. Veg Hist Archaeobot 17:63–71
9. McGovern PE (2019) Ancient wine: the search for the origins of viniculture (Princeton Science library 76). Princeton University Press, Princeton
10. Kalantaryan I (2005) Newly discovered rock-cut mausoleum of Agarak. Cult Anc Armen 13:154–160
11. Wright JC (2004) A survey of evidence for feasting in Mycenaean society. Hesperia J Am Sch Class Stud Athens 73:133–178

12. Donahue JF (2015) Food and drink in antiquity: readings from the Graeco-Roman World: a sourcebook. Bloomsbury Academic, London
13. Jouanna J (2012) Wine and medicine in ancient Greece. In: van der Eijk P, Allies N (eds) Greek medicine from Hippocrates to Galen: selected papers. Brill, Leiden, pp 173–193
14. Renfrew JM (2003) Palaeoethnobotanical finds of Vitis from Greece. In: McGovern PE, Fleming SJ, Katz SH (eds) The origins and ancient history of wine. Routledge, London, pp 255–267
15. Redford DB (1992) Egypt, Canaan, and Israel in ancient times. Princeton University Press, Princeton
16. Boso S, Gago P, Santiago JL, Teira-Brión A, Martín-Seijo M, Rey-Castiñeira J, Ocete R, Ocete C, Martínez MC (2020) Morphometric comparison of current, Roman-era and medieval Vitis seeds from the north-west of Spain. Aust J Grape Wine Res 26:300–309
17. Van Limbergen D (2020) Agroforestry and the reappraisal of Roman viticulture. Glob Environ 13:432–450
18. Cantos M, Arroyo-García R, García JL, Lara M, Morales R, López MÁ, Gallardo A, Ocete CA, Rodríguez Á, Valle JM et al (2017) Current distribution and characterization of the wild grapevine populations in Andalusia (Spain). Comptes Rendus Biol 340:164–177
19. World wine production outlook, OIV first estimates, 31.10.2022. International Organisation of Vine and Wine

Chemistry of Wines

2

"A glass of wine...," there may not be a person in the world who has not tasted a glass of wine. This magical potion that starts from the grape becomes juice and without any absolute intervention turns into wine. An art that became a science. A science that is constantly evolving. Millions of people work so that we can enjoy a glass of wine. So simple yet so complex at the same time. We will try to analyze this complexity in this chapter. Hundreds of chemicals interact and create new ones. Numerous biochemical reactions create this delectable result called wine. Knowledge of the chemical composition of wine is a prerequisite for the professional oenologist or chemist. Without the absolute knowledge of it, we cannot create a perfect wine. Imagine a painter who doesn't know what brush to use or a musician who doesn't know the notes. Absolute knowledge of wine chemistry is what will lead us to the right decisions at the right time. Through the appropriate chemical analyzes, we will be able to make the important decisions, apply the appropriate winemaking techniques, and save money and time during production.

2.1 Alcohols

In the table below, we can see which alcohols originating from plants and yeast (Table 2.1) [1].

2.1.1 Methyl Alcohol

Methyl alcohol content in wines is estimated at around 30–350 mg/L and is not formed by alcoholic fermentation. The structure of the molecule of methanol is CH_3–O. Its presence is a result exclusively from enzymic hydrolysis of the methoxyl groups of the pectins during fermentation (Fig. 2.1) [1].

G. Z. Kyzas, F. Papageorgiou, *Oenology in Practice*,
https://doi.org/10.1007/978-3-031-85531-3_2

Table 2.1 Formulas of various alcohols originating from plants and yeasts

Name	Formula
Methanol	$H–CH_2OH$
Ethanol	$CH_3–CH_2–OH$
Propanol-1	$CH_3–CH_2–CH_2OH$
Propanol-2	$CH_3–CHOH–CH_3$
Butanol-1	$CH_3–CH_2–CH_2–CH_2OH$
Methyl-2-propanol-1	$(CH_3)_2–CH–CH_2OH$
Methyl-2-propanol-2	$(CH_3)_2–COH–CH_3$
Butanol-2	$CH_3–CH_2–CHOH–CH_3$
Butanediol-2,3	$CH_3–CHOH–CHOH–CH_3$
Pentanol-1	$CH_3–CH_2–CH_2–CH_2–CH_2OH$
Pentanol-2	$CH_3–CH_2–CH_2–CHOH–CH_3$
Pentanol-3	$CH_3–CH_2–CHOH–CH_2–CH_3$
Methyl-3-butanol-1	$(CH_3)_2–CH–CH_2–CH_2OH$
Methyl-2-butanol-1	$CH_3–CH_2–CH(CH_3)–CH_2OH$
Methyl-3-butanol-1	$(CH_3)_2–CH–CHOH–CH_3$
Hexanol-1	$CH_3–(CH_2)_4–CH_2OH$
Hexanol-2	$CH_3–(CH_2)_3–CHOH–CH_3$
Heptanol-1	$CH_3–(CH_2)_5–CH_2OH$
Heptanol-1	$CH_3–(CH_2)_4–CHOH–CH_3$
Octanol-1	$CH_3–(CH_2)_6–CH_2OH$
Octanol-2	$CH_3–(CH_2)_5–CHOH–CH_3$
Nonanol-1	$CH_3–(CH_2)_7–CH_2OH$
Nonanol-2	$CH_3–(CH_2)_6–CHOH–CH_3$
Decanol-1	$CH_3–(CH_2)_8–CH_2OH$
Phenyl-2-ethanol	$\Phi–CH_2–CH_2OH$
Tyrosol	$HO–\Phi–CH_2–CH_2OH$
Octene-1-ol-3	$CH3–(CH_2)_4–CHOH–CH{=}CH_2$

Φ = benzene cycle

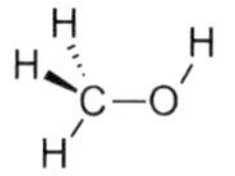

Fig. 2.1 Structure of methyl ethanol molecule

Grapes have relatively low pectin content, resulting in very little production of methyl alcohol. Research studies have shown that red wines have a higher concentration than rose wines and even more than white wines. This has a logical explanation as the concentration of methanol is proportional to the soaking time of the grape skins which are high in pectins.

Another important piece of information is that the wines are made of hybrids grapes have higher methanol content than wines are made of *Vitis vinifera* grapes. That happens because hybrid grapes have high pectin content. Increase in methanol content caused by the use of pectolytic enzymes at the beginning of winemaking. Methanol is toxic. 100 mL of methanol is enough to cause death.

2.1.2 Ethyl Alcohol

Ethyl alcohol is the main product of sugar metabolism by yeasts. Usually the ethanol content of wines ranges from 10% to 16% of its volume. Alcohol together with reducing sugars and glycerol are the sweet components of wines and moderate the sour taste of acids and the bitterness of phenolic compounds. In addition to the complex and special taste that alcohol imparts, its aroma forms the basis for the aroma and bouquet of wines. Finally, alcohol facilitates the preservation of wines thanks to its antiseptic nature. We need 18 g/L of sugar is required to produce 1% of ethanol during alcoholic fermentation. This means that a must containing 226 g/L will produce a wine with 12.6% alcohol by volume.

There are many who have equated the quality of the wine with the alcohol it contains. In some cases, the price is even affected by the alcohol content of the wine. In reality, however, the alcohol content of a wine is purely related to the type of wine that one is going to vinify. What do we mean by this, even if we want to make a wine intended for aging, surely in this case, we should make a wine with a high alcoholic strength. This is because ethanol has disinfectant properties that are particularly valuable in aging wine. The combination of ethanol and acidity makes it possible to keep wine for a long time without any noticeable spoilage. Suppose now that we want to vinify an "everyday wine" intended for immediate consumption. In this case, we want a wine of either low or medium alcohol content because this wine will be consumed immediately.

In chemical terms, ethanol is a primary alcohol specifically is a 2-carbon alcohol. The structure of the molecule of ethanol is CH_3–CH_2–OH (an ethyl group linked to a hydroxyl group), which indicates that the carbon of a methyl group (CH_3–) is attached to the carbon of a methylene group (–CH_2–), which is attached to the oxygen of a hydroxyl group (–OH). It is a constitutional isomer of dimethyl ether.

Ethanol is a volatile, colorless liquid that has a slight odor. It burns with a smokeless blue flame that is not always visible in normal light. The physical properties of ethanol stem primarily from the presence of its hydroxyl group and the shortness of its carbon chain. Ethanol's hydroxyl group is able to participate in hydrogen bonding, rendering it more viscous and less volatile than less polar organic compounds of similar molecular weight, such as propane (Fig. 2.2).

2.1.3 Higher Alcohols

Monoalcohols that contain more than two carbon atoms in their molecule belong to this group of alcohols. Higher alcohols at low concentrations—below 300 mg/L—contribute to wine's aromatic complexity.

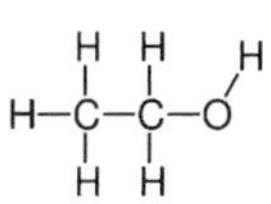

Fig. 2.2 Structure of ethyl ethanol molecule

Higher alcohols are secondary products of alcoholic fermentation with the exception of hexanol-1, which is derived directly from the grape. Higher alcohols come from ketone acids which are formed by completely different mechanisms, both from sugars and from amino acids.

Higher alcohols formation depends on several factors such as the chemical composition of the must, the aeration conditions of the must during alcoholic fermentation, the type of yeasts we use, the fermentation temperature, the active acidity (pH), and the total acidity of the must. To be more specific, the significant aeration of grape must during alcoholic fermentation increases the amount of higher alcohols produced. Research studies have shown that the production of higher alcohols is proportional to the total alcoholic strength. If we wish to increase the amount of higher alcohols in our produced wine, then we should carry out the alcoholic fermentation at as high an active acidity (pH) as possible and as low a temperature as possible [1].

2.1.4 Other Alcohols

This separate category includes alcohols whose composition is due exclusively to the raw material, specifically grapes. The first group of this category is C_6 alcohols, hexanols, and hexenols. Their production gives to wine an herbaceous smell. A second category is octen-1-ol-3. A wine with a high content of this alcohol gives a characteristic mushroom aroma. Botrytis cinerea is responsible for the production of this specific alcohol. Finally, we have terpenols. Terpenols is the main component of Muscat wines.

2.1.5 Polyalcohols

Sugar alcohols (also called polyhydric alcohols, polyalcohols, alditols, or glycitols) are organic compounds, typically derived from sugars, containing one hydroxyl group (−OH) attached to each carbon atom. Accumulation of hydroxyl radicals in a compound raises the boiling point considerably due to the large number of hydrogen bonds as well as increasing its viscosity. They are white, water-soluble solids that can occur naturally or be produced industrially by hydrogenating sugars.

2.1.5.1 C_3 Polyols, Glycerol, $CH_2OH{-}CHOH{-}CH_2OH$

Glycerol, besides water and ethanol, is usually the chemical compound with the highest concentration in wine. Its concentration in wines ranges from 5 to 20 g/L. Glycerol is chemically stable, but not biologically stable. It can be affected by lactic acid bacteria with the formation of lactic and acetic acid but also acrolein which is responsible for the bitter taste of these affected wines. Glycerol is created at the beginning of alcoholic fermentation from the first 50 g of fermented sugars more than half the amount of glycerol contained in wines is produced (Fig. 2.3).

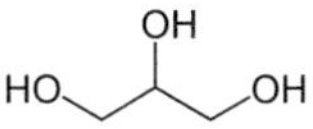

Fig. 2.3 Structure of glycerol molecule

The total glycerol content depends on the initial amount of sugars, the type of yeasts, and the conditions under which the alcoholic fermentation is carried out (temperature, oxygen, and amount of sulfite anhydride). Specifically, it has been established that its formation is favored by relatively higher concentrations of tartaric acid, by the addition of sulfuric anhydride and by low-fermentation temperatures.

At this point it is worth noting that wines obtained from grapes that have been affected by the fungus botrytis cinerea have increased glycerol content. That is why, during the production of vins liquoreux, the greatest possible concentration of glycerol is sought and is a great advantage.

Glycerol is of great technological interest in the wine industry. It has a sweet taste and gives dry wines a special sense of sweetness. Another characteristic of glycerol is its increased viscosity which, in combination with the alcohol content of the wines, gives the wines a characteristic body (how heavy and rich a wine tastes).

2.1.5.2 C4 Polyols, 2,3-Butanediol, CH_3–CHOH–CHOH–CH_3

The content of the wines in butanediol-2,3 ranges from 0.3 to 1.4 g/L. It is produced during alcoholic fermentation. Specifically, it comes from the reduction of acetoin (3-hydroxybutanone or acetyl methyl carbinol), which in turn results from the condensation of two acetaldehyde molecules, one of which is active. 2,3-butanediol is probably also formed by malolactic fermentation. This chemical compound has a little odor and its flavor is slightly sweet and bitter at the same time (Fig. 2.4).

The 2,3-butanediol content of wines is a key criterion for determining whether wines have been fortified (alcohol added). Also very high levels of it show suspicion of strengthening the alcoholic strength with the addition of sugar.

Research studies have shown that 2,3-butanediol is produced in greater quantities when alcoholic fermentation is done by the type of yeast *Schizosaccharomyces* pomb compared to all other yeasts. Also, its composition is favored in cases where the alcoholic fermentation takes place at high temperatures. In white vinification, we have about 20–25% more production of 2,3-butanediol when the must has been de-slugged in time. In red winemaking, if we heat the must and the grapes, then we have a reduction of 2,3-butanediol of the order of 15%.

The most significant technological role of 2,3-butandiol is in maintaining an oxidation–reduction balance with acetoin and diacetyl. Acetoin has a slight milky odor at concentration of 10 mg/L and diacetyl has a pleasant odor of butter and hazelnuts at concentrations about 2 mg/L.

In C4 polyols, we have another chemical compound called erythritol. Erythritol CH_2OH–$(CHOH)_2$–CH_2OH has four alcohol functions and the properties are not known. Concentration of this chemical compound range from 30 to 200 mg/L and is formed by yeasts.

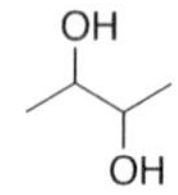

Fig. 2.4 Structure of 2,3-butanediol, molecule

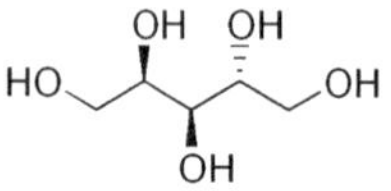

Fig. 2.5 Structure of Arabitol molecule

2.1.5.3 C5 Polyol, Arabitol $CH_2OH–(CHOH)_3–CH_2OH$

Arabitol can be found as L-arabitol and D-arabitol. We found this chemical compound in small quantities in wines range from 25 to 350 mg/L. These isomers have different production pathways. D-arabitol is obtained from the hydrogenation of xylulose, by the enzyme D-arabitol dehydrogenase. *Candida* sp., *Hansenula* sp. [2], and Pichia [3] are some examples of yeast producers of D-arabitol. In contrast, L-arabitol is obtained by direct hydrogenation of L-arabinose catalyzed by L-arabinose reductase, which commonly occurs in yeast such as Candida, Kluyveromyces, and Pichia [4]. Small quantities can also be produced by lactic bacteria and larger quantities by botrytis cinerea (Fig. 2.5).

2.1.5.4 C6 Polyols, Mannitol, $CH_2OH–(CHOH)_4–CH_2OH$

Mannitol is derived from reduction of the C1 on mannose. In wines, mannitol results from the reduction of fructose by some lactic bacteria. It has four asymmetric carbon atoms and has a sweet taste. Its usual content in wines is about 40 mg/L. Sometimes the reduction of fructose by lactic bacteria is intense. This happens in cases where, for various reasons, we have an interruption of the alcoholic fermentation. In these cases, the mannitol content reaches 10 g/L. When this happens, we are talking about the case of manitic fermentation. The formation of mannitol is always accompanied by the formation of erythritol and arabitol (Fig. 2.6).

2.1.5.5 C6 Polyols, Sorbitol, $CH_2OH–(CHOH)_4–CH_2OH$

Sorbitol is an isomer of mannitol and its content in wines is around 20–30 mg/L and is totally absent by healthy grapes. Large quantities of sorbitol indicate that wine must has been mixed with must of other fruits. Apples and pears have high sorbitol content (Fig. 2.7).

2.1.5.6 C6 Polyols, Meso-Inositol, $(CHOH)_6$

Meso-inositol is a chemical component of grapes and wine. Its chemical form incudes a cyclic polyols with six-carbon atoms, each one carrying a hydroxyl radical. Its content in must and wines is about 0.5 g/L. It has a sweet taste, like sugar and properties that characterize vitamins (Fig. 2.8).

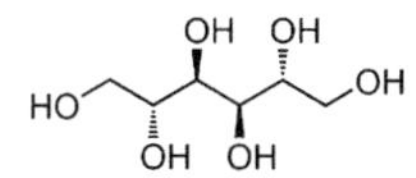

Fig. 2.6 Structure of Mannitol molecule

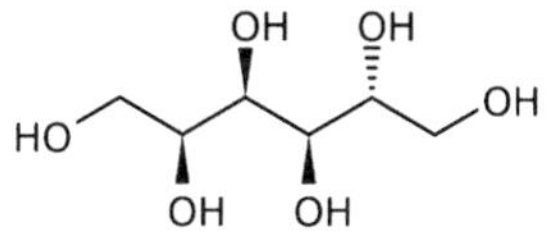

Fig. 2.7 Structure of Sorbitol molecule

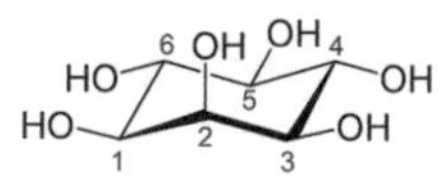

Fig. 2.8 Structure of Meso-inositol molecule

2.2 Acids in Wine

2.2.1 Introduction

Must and wine contain inorganic and organic acids as well as a large amount of bases that neutralize all of the inorganic acids as well as part of the organic ones. A part of the organic acids remains free. The acidity of the must is also due to this. Their preservative properties enhance wines' microbiological and physicochemical stability. They are responsible for the acidic taste of the wines and apart from the beneficial properties mentioned before, they are also responsible for maintaining the color of the wines.

2.2.2 Organic Acids

The organic acids included in wines have a double origin. A part of them comes from the grape and the rest is formed during the various fermentations we have during the winemaking process. The main acids of these two categories are as follows.

2.2.2.1 Acids Derived from Grapes

Below we will deal with the organic acids found in wine which come directly from the raw material. It is very important in our science to know their origin because each of them has a special technological interest and this knowledge will help us to choose the appropriate winemaking protocol that we will see in the next chapter. However, the protocols in our science must always be adapted to the raw material we receive for winemaking, so it is very useful to know the chemical composition and the origin of the chemical elements included in it, in order to have the best possible result. The order in which the organic acids are presented is not random since it is decreasing in relation to the concentration of the acids in musts and wines.

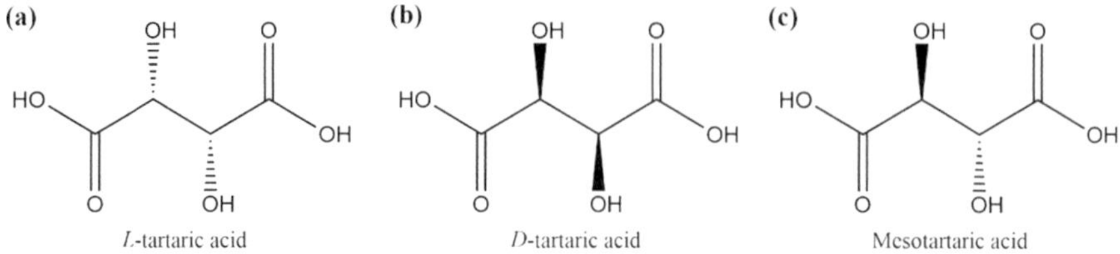

Fig. 2.9 Optical isomers of tartaric acid. (**a**) L-tartaric acid, (**b**) D-tartaric acid, (**c**) Mesotartaric acid

2.2.2.1.1 Tartaric Acid

Tartaric acid is considered the specialty acid of the grape. The natural tartaric acid is D-tartaric. In nature in general it is very little widespread but is one of the most prevalent acids in grapes and must. Concentrations in unripe grapes may be as high as 15 g/L. During their ripening, part of the tartaric acid is lost and the rest is dissolved in the new amounts of water that swell the grapes. This has the result that its content in ripe grapes drops to 7–8 g/L. The final content of tartaric acid in wines is affected by several factors, resulting in an even lower concentration of 1.5–3.0 g/L (Fig. 2.9).

These factors are:

- In general, the annual temperature fluctuations and in particular the weather conditions during ripening and even more specifically a few days before the harvest.
- The vine variety as well as the type of grafted clone.
- The type and composition of the soil where the vineyard is located.
- The cultivation cares done during the year.
- The sugars concentrations of grapes and the total alcohol after the alcoholic fermentation.
- The content of malic acid in must.
- The content of lactic acid in wine.
- The content of inorganic anions.
- The temperature of the environment during the maturation of the wine.
- The type of wine we plan to vinify.
- Possible attack of tartaric acid by certain lactic bacteria leading to wine disease which reduces the tartaric acid content to zero.

2.2.2.1.2 Malic Acid

Unlike tartaric acid, malic acid is very widespread in nature. Natural malic acid is the l isomer. In grapes, its content changes as the fruit matures. In the first stage, when we have green grapes, the content of malic acid can reach up to 25 g/L. This concentration decreases drastically at the stage when the grapes are ripe and ready for harvest and reaches an average of 5 g/L (Fig. 2.10).

Malic acid is relatively easily attacked by yeasts and bacteria. One such case is the one where a type of yeast and specifically *Schizosaccharomyces* pombe convert malic acid into alcohol. A second and more common case in the wine industry is the conversion of malic acid to lactic acid by certain lactic acid bacteria. The latter case,

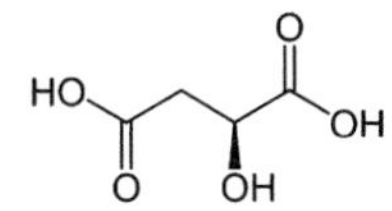

Fig. 2.10 Fischer projection of L(−) Malic acid

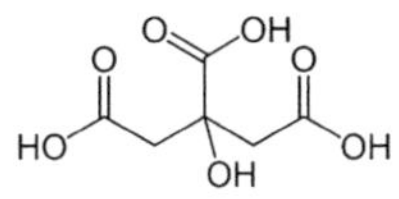

Fig. 2.11 Fischer projection of citric acid

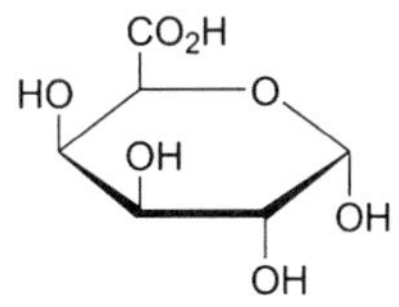

Fig. 2.12 Fischer projection of D-galacturonic acid

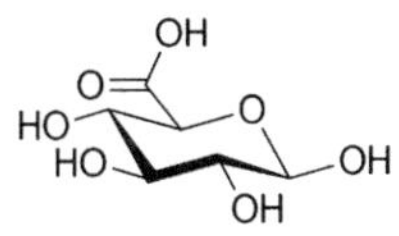

Fig. 2.13 Fischer projection of D-glucuronic acid

also known as malolactic fermentation, is usually desired in the wine industry as it improves the organoleptic characteristics of the wine.

2.2.2.1.3 Citric Acid

Citric acid is very widespread in nature, especially in lemons but also in citrus fruits in general. Its content in grapes is about 0.5 g/L. An important detail about citric acid is the fact that its content in wine comes entirely from grapes. It is worth noting that the addition of citric acid to increase total acidity is allowed and indicated but only in wines and not in must. Citric acid is quite unstable toward lactic acid bacteria, so interventions with it, especially in red wines, require special attention because in the event of its decomposition, we have an increase in volatile acidity (Fig. 2.11).

2.2.2.1.4 Galacturonic Acid

Galacturonic acid results from the hydrolysis of grape pectin materials. During the same process, in addition to galacturonic acid, we also have the production of methanol. This process begins (mainly in red vinification) at the stage of pressing or breaking the grapes. That is why red wines have almost twice (and more) the content of galacturonic acid compared to white vinification, where we do not have coexistence of the grapes with the must (Fig. 2.12).

2.2.2.1.5 D-Glucuronic Acid

Glucuronic acid and galacturonic acid belong to the category of uronic acids. The content of this specific acid is found in amounts up to 1.3 g/L and comes from the enzymatic oxidation of glucose (Fig. 2.13).

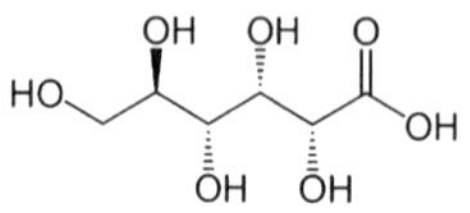

Fig. 2.14 Fischer projection gluconic acid

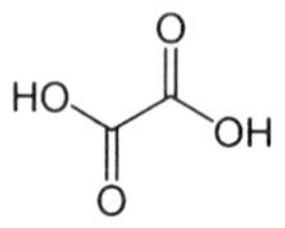

Fig. 2.15 Fischer projection of oxalic acid

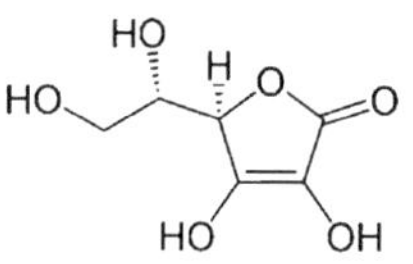

Fig. 2.16 Fischer projection of ascorbic acid

2.2.2.1.6 Gluconic Acid

The content of gluconic acid in wines can also be characterized as an indicator of the healthy state of the grapes. Specific research has shown that in wines derived from healthy grapes the concentration of gluconic acid is less than 0.15 g/L. In cases where the grapes have been affected by fungal diseases, the concentration of gluconic acid can reach up to 2.5 g/L. The origin of gluconic acid comes from the oxidation of glucose and specifically of one of its aldehydic groups (Fig. 2.14).

2.2.2.1.7 Oxalic Acid

Oxalic acid is a natural acid found in grape must. However, in cases where we have oxidation of tartaric acid (which is one of the most stable acids in must), an excess amount of oxalic acid can be formed. Its concentration in wines is less than 0.07 g/L. In the science of oenology, this specific acid, despite its low concentration, presents an important property. Specifically, oxalic acid is bound to trivalent iron ions, when the wine is in a reducing environment, it is known that trivalent iron ions is reduced to divalent iron ions. This results in the oxalic acid being found free in the wine and reacting with the calcium, forming calcium oxalate (Fig. 2.15).

2.2.2.1.8 Ascorbic Acid

Ascorbic acid or otherwise vitamin c is contained in the grape must before the start of the alcoholic fermentation in amounts that reach 100 mg/L. This available quantity is consumed by the yeasts and thus in the wine the available quantity is almost zero. It is an organic acid of great technological interest, its antioxidant action it protects the wine from future oxidation and cloudiness. Its addition is allowed in the wine industry, but we must consult the legislation to know the maximum amount allowed. It is worth noting that the addition of ascorbic acid affects the measurement of total and free sulfuric anhydride (Fig. 2.16).

2.2.2.2 Acids Resulting from Various Fermentations

In this section, we will deal with the organic acids resulting from the various fermentations that take place during the conversion of must into wine. In simpler words, the organic acids to be analyzed did not pre-exist in the fruit or must but were created after a series of chemical reactions. This is a class of acids with a very important technological interest and this is the reason why we analyze each one separately.

2.2.2.2.1 Succinic Acid

Succinic acid is produced during alcoholic fermentation. Its content in wines ranges from 0.0 to 2.6 g/L. This particular acid is very resistant to attacks by microorganisms and bacteria and its taste is special. Succinic acid dissolved in water has an unusual salty and bitter taste. When compared to solutions of tartaric acid, tasters reported that succinic acid solutions in water (0.5, 1.0, and 2.0 g/L) were unpleasant and Succinic acid dissolved in water has an unusual salty and bitter taste [5]. When compared to solutions of tartaric acid, tasters reported that succinic acid solutions in water (0.5, 1.0, and 2.0 g/L) were unpleasant and indicated that the unusual taste lingered after expectorating cated that the unusual taste lingered after expectorating The increase in succinic acid during alcoholic fermentation can lead to an increase in the wine's total acidity. A survey of 138 Australian wines—93 red wines and 45 white wines—showed that succinic acid concentration in finished red wines ranged from 0 to 2.6 g/L with a mean concentration of 1.2 g/L. In the white wines, the succinic acid concentration ranged from 0.1 to 1.6 g/L with a mean concentration of 0.6 g/L. In general, the concentration of succinic acid in white wines is about half the concentration of succinic acid found in red wines [6] (Fig. 2.17).

2.2.2.2.2 Lactic Acid

Lactic acid and its two isomers, in contrast to the malic acid analyzed above, is an organic acid that is only detected in wines and not at all in musts. It is very interesting to know the ratio of the isomers of lactic acid as this is how we understand the mechanism of its creation. Lactic acid has three different mechanisms that are created in wine. The first is from yeasts during alcoholic fermentation, the second is from lactic bacteria with malolactic fermentation, and the third again from lactic bacteria that attack sugars and tartaric acid. In the first case, we have mainly production of the D(−) isomer, in the second case, we have exclusively production of the L(−) isomer, and in the third case, we have simultaneous production of both isomers. Taking into account all the above, we calculate that the content of lactic acid in wines can be from 0.1 up to 10 g/L, depending on the mechanism of its creation. Lactic acid is often associated with "milky" flavors in wine and is the primary acid of yogurt and sauerkraut. Lactic acid adding complexity and softening the harshness of malic acidity, but it can generate off flavors and turbidity in others (Fig. 2.18).

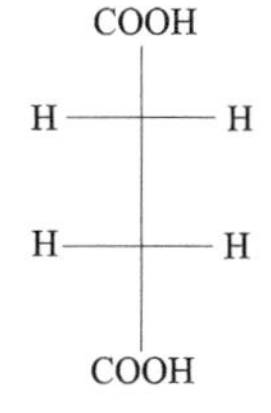

Fig. 2.17 Fischer projection of succinic acid

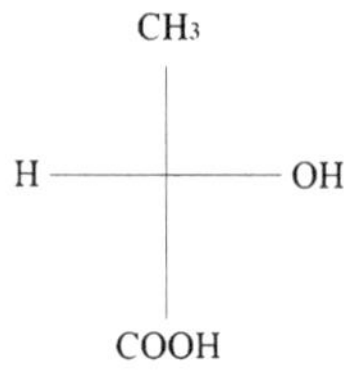

(1)

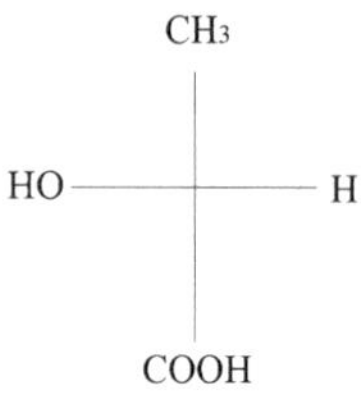

(2)

Fig. 2.18 Fischer projection of D(−) Lactic acid (1) and L(+) Lactic acid

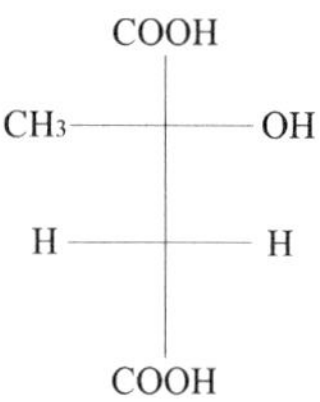

Fig. 2.19 Fischer projection of citramalic acid

2.2.2.2.3 Citramalic Acid

This specific acid is produced as secondary components of alcoholic fermentation in concentrations between 0.1 and 0.25 g/L. The presence of citramalic acid characterizes sweet fortified wines (Fig. 2.19).

2.2.3 Volatile Acids Resulting from Various Fermentations

The acids that will be analyzed next belong to the category of acids that arise during the various fermentations that take place during the process of converting grape juice into wine. Their difference with previous acids of the same class is that the resulting acids are volatile acids. Of course, these acids are also counted in the sum of the total acidity, but the acids of the volatile acidity in the science of oenology must be studied separately because of the great technological interest they present.

When we hear that a wine has volatile acidity, we automatically conclude that it is a problematic wine. But at this point, we must clarify that all wines have volatile acidity since a part of the volatile acidity comes from alcoholic fermentation and another part comes from malolactic fermentation. The problem starts when volatile acidity is created by unwanted fermentations such as oxidation of alcohol through the chemical or enzymatic pathway or results from bacterial infestation.

2.2.3.1 Acetic Acid

Acetic acid is one of the most important acids in the wine industry. In itself it is an indicator of the quality of the wines produced and it is the largest percentage of volatile acidity at a percentage of up to 95%. It is an acid that we don't want in our wines since its main production route comes from bacterial contamination of the wine's components. Acetic acid bacteria require oxygen to grow and proliferate. Like many other micro-organisms affiliated with wine production, acetic acid bacteria can be managed with proper sulfur dioxide treatments, adequate temperature control, thorough sanitation practices, and appropriate oxygen management

Fig. 2.20 Fischer projection of acetic acid

strategies depending on the wine style. Reducing oxygen is a good way to minimize acetic acid management growth due to the bacteria's dependence on oxygen. There is strict legislation on the content of your wines that, depending on the country of production, we must consult. Its presence significantly degrades the organoleptic characteristics of a wine since a high concentration gives it the smell of vinegar. It is very important with our interventions to keep the concentration of this particular acid as low as possible. Its concentration in quality wines must be less than 0.5 g/L and in inferior wines below 0.7 g/L. Concentrations greater than 0.7 g/L but at the same time below the legal limit should lead us to quick decisions about the future of our wine. An important role in these decisions is played by the concentration of ethyl acetate since this particular chemical compound is responsible for vinegar odors and not the acetic acid itself. It is important to distinguish the odors between acetic acid and ethyl acetate. Specifically, acetic acid smells like vinegar and ethyl acetate smells like nail polish or some nail polish removers (Fig. 2.20).

2.2.4 Different Types of Acidity

It is a fact that during the production of wine, the oenologist must make quick and correct decisions. This combined with the chemical complexity of wine makes the job even more difficult. In many cases, there is neither the time nor the required laboratory equipment for the quantitative and qualitative detection of all these acids analyzed so far. For this reason, a winemaker should focus on total acidity, active acidity, volatile acidity, and fixed acidity. These measurements are quick, easy, and lead us to safe conclusions about the quality state of the wine. Knowing these we can make very important decisions about winemaking and intervene accordingly.

2.2.4.1 Total Acidity

Total acidity expresses the set of free carboxyl groups of the acids, whether they are dissociated or not. Total acidity (T.A) of must or wine takes into account all types of acids (inorganic, organic).

2.2.4.2 Active Acidity (pH)

Active acidity expresses the total number of carboxyl groups that are in dissociation and corresponds to the total number of hydrogen ions (H^+). It is a fact that it is the concept of pH it is a bit abstract. But it is a very important measurement that is very easily done with the use of a pH meter. The values of pH have a range between 2.8 and 4.0. Low values are good for many reasons, but the most basic reason is the fact that, at low pH micro-organisms find it difficult to survive and multiply. Equally important is that in low prices of pH favor the use of sulfuric anhydride.

2.2.4.3 Volatile Acidity

Volatile acidity has been analyzed previously, but if we wanted to summarize in one paragraph its meaning and importance in wine, we could say that volatile acidity in wine consists of the free and conjugated forms of volatile acids. It is part of the total acidity, but its importance forces us to calculate it separately.

2.2.4.4 Fixed Acidity

Fixed acidity content of a wine is obtained by subtracting volatile acidity from total acidity.

2.3 Carbohydrates

Carbohydrates also known as sugars are produced by photosynthesis in vine leaves. Photosynthesis is the process in plants and certain other organisms that use the energy from the sun to convert carbon dioxide and water into glucose (a sugar) and oxygen. The equation for photosynthesis is: $6CO_2 + 6H_2O \rightarrow C_6H_{12}O_6 + 6O_2$.

This means that the reactants, six-carbon dioxide molecules and six water molecules, are converted by light energy captured by chlorophyll into a sugar molecule and six oxygen molecules.

Sugars are compounds consisting of a carbon chain which contains alcohol groups and in addition an aldehyde or a ketone group. Their general formula is $C_V(H_2O)_V$. Sugars consist of polyfunctional molecules, capable of participating in a large number of chemical, biochemical and metabolic reactions. Focusing on the properties of sugars that concern the science of oenology, we notice that they are the precursors of organic acids. During alcoholic fermentation, glucose and fructose generate ethanol and other by-products. Furthermore, glucose and fructose can be attacked by lactic bacteria to produce lactic acid, sometimes acetic acid produced by the same bacteria. The sugars of must and wine are divided into reducing and non-reducing. More specifically:

- Reducing sugars (reduce Fehling's solution).
  - Fermentable sugars.
  - Non-fermentable sugars.
- Non-reducing sugars (does not reduce Fehling's solution).
  - Saccharose, raffinose, stachyose, and trehalose.

2.3.1 Reducing Sugars

Reducing sugars are frequently used in winemaking process. Reducing sugars have an aldehyde or ketone function that reduces the alkaline cupric solutions used to assay them and they consist of hexoses and pentoses. The fermentable sugars of must or wine are composed of hexoses and specifically are the D(+) Glucose, D(−) Fructose, and D(−) Galactose. Reducing sugars include pentoses too but they are not fermentable from yeasts. Both fermentable and non-fermentable reducing sugars have some important properties. First of all they reduce Fehling's Solution, second they are biologically unstable and they bind the disulfide anhydride. It is important to determine the reducing sugars because depending on their content in the final wines there is a legislative categorization of them.

Dry wines: below or equal to 2 g/L.
Semi dry wines: between 2 and 18 g/L.
Semi sweet wines: between 18 and 40 g/L.
Sweet wines: over 40 g/L.

2.3.1.1 D(+) Glucose

The name glucose is derived from Ancient Greek γλεῦκος (gleûkos, "wine, must"), from γλυκύς (glykýs, "sweet") [7]. The suffix "-ose" is a chemical classifier denoting a sugar. Glucose is also called dextrose because it has the property of turning the plane of incident light to the right (α) = +52.5o. It forms white or colorless solids that are highly soluble in water and acetic acid but poorly soluble in methanol and ethanol (Fig. 2.21).

2.3.1.2 D(−) Fructose

The word "fructose" was coined in 1857 from the Latin for fructus (fruit) and the generic chemical suffix for sugars, -ose [8]. Fructose, also called levulose and has the property of turning the plane of polarized light (a)-92.5o to the left (Fig. 2.22).

At this point, we will deal more extensively with the two sugars that play a leading role in the winemaking process. We are talking about Glucose (G) and Fructose (F). As mentioned in the introduction, the various sugars are produced during the

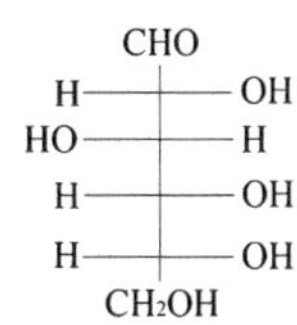

Fig. 2.21 Fischer projection of D(+) Glucose

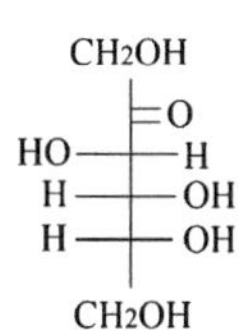

Fig. 2.22 Fischer projection of D(−) fructose

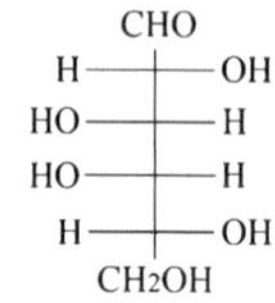

Fig. 2.23 Fischer projection of D-Galactose

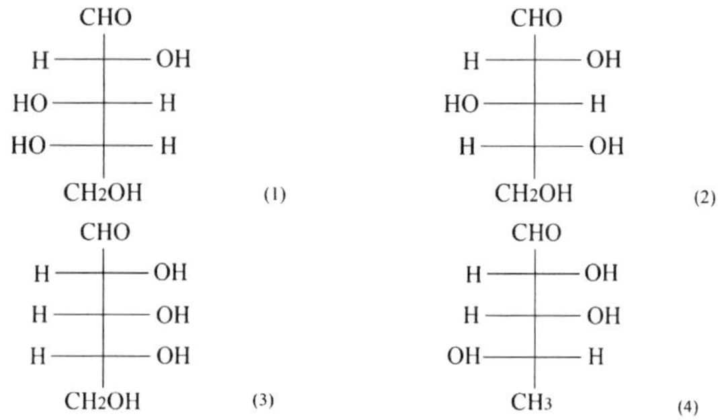

Fig. 2.24 Fischer projection of L-arabinose (1), D-xylose (2), D-ribose (3), L-rhamnose (4)

process of photosynthesis and the grape fruit is their storehouse. To make the comparison of these two sugars (G) and (F) easier to understand, we will use the ratio (G)/(F). The grapes follow a natural maturation that goes through several stages. At the stage when the fruit is green, there is more glucose (G) than fructose (F), a common ratio (G)/(F) is about 1.5. With the alternation of ripening stages, this ratio tends to become 1. This is because during the ripening process, the proportion of fructose increases. At the full maturity of the grape, we have and the vintage, this ratio (depending on the type of wine we are planning to vinify) can fall below 1. During alcoholic fermentation and because yeasts ferment glucose more easily than fructose, the (G)/(F) ratio becomes very small compared to their initial ratio in grape must. For the same reason, if we do a qualitative and quantitative analysis of the hexoses present in dry wines, it is most likely that fructose prevails. Another very important comparative feature of the two protagonist hexoses is the differentiation of sweetness, specifically if we wanted to quantify sweetness and if Saccharose has a rating of 1.0 fructose rates 1.73 and glucose 0.74. The overall glucose and fructose concentration in ripe grape juice is between 130 and 250 g/L.

2.3.1.3 D-Galactose

D-Galactose is a hexose that is fermented by a small number of yeast cells is also known as brain sugar since it is a component of glycoproteins (oligosaccharide-protein compounds) found in nerve tissue (Fig. 2.23).

2.3.1.4 Non-fermentable Sugars

The non-fermentable reducing sugars of must and wine are composed in their majority of pentoses. The main pentoses identified in grapes and wines are L-arabinose, D-xylose, D-ribose, and L-rhamnose (Fig. 2.24).

L-arabinose, D-xylose, D-ribose, and L-rhamnose as we can see in Fischer projection above is an aldopentose—a monosaccharide containing five carbon atoms, and including an aldehyde (–CHO) functional group. The total concentration of pentoses in wines usually does not exceed 2 g/L. Higher concentrations than usual are usually accompanied by mixing of grape must with other fruit must.

In general, red wines are richer in pentoses than white wines due to the fact that this specific category of sugars is contained in a greater proportion in the skins and skins of the grapes than in their flesh. As previously mentioned, pentoses do not belong to fermentable sugars, but they can be affected by lactic acid bacteria and we can have an increase in the volatile acidity of wines.

Besides pentoses, there are some other reducing sugars in grape and wines that are non-fermentable. Melibiose is one of them. Melibiose is a reducing non-fermentable disaccharide formed between galactose and glucose. It can be formed by invertase-mediated hydrolysis of raffinose, which produces melibiose and fructose. Melibiose can be broken down into its component saccharides, glucose, and galactose, by the enzyme alpha-galactosidase, such as MEL1 from *Saccharomyces pastorianus* (lager yeast).

Maltose is another non-fermentable reducing sugar. Maltose also known as maltobiose or malt sugar is a disaccharide formed from 2 units of glucose. Maltose can be broken down to glucose by the maltase enzyme, which catalyzes the hydrolysis of the glycosidic bond.

Lactose at the same sugar category as maltose and melibiose consists of one molecule of glucose and one molecule of galactose. Lactose is hydrolyzed to glucose and galactose, isomerized in alkaline solution to lactulose, and catalytically hydrogenated to the corresponding polyhydric alcohol, lactitol. Melibiose cannot be used by *Saccharomyces cerevisiae*.

2.3.2 Non-reducing Sugars

In the category of non-reducing sugars and non-fermentable sugars with the disaccharide Saccharose as the protagonist, there are also raffinose, trahyose, and trehalose. Saccharose or sucrose is a non-reducing disaccharide and consists of one glucose molecule and one fructose molecule. Sucrose is a non-reducing disaccharide and consists of one glucose molecule and one fructose molecule. Its content in ripe grapes ranges from 2 to 5 g/L, although its content may occasionally be slightly higher. In the disaccharide form, it is not fermentable. This changes in the course of alcoholic fermentation as with the help of the enzyme invertase, the sucrose molecule is hydrolyzed into D-glucose and D-fructose. In addition to sucrose, there are three more sugars in the grape in the same category of non-reducing and non-fermentable sugars. These are raffinose, stachyose, and trehalose. Raffinose is a trisaccharide composed of galactose, glucose, and fructose. Stachyose is a tetrasaccharide consisting of two α-D-galactose units, one is α-D-glucose unit and one is β-D-fructose. Stachyose is less sweet than sucrose, at about 28% on a weight basis. Trehalose is a disaccharide consisting of two molecules of glucose. It is also known as mycose or tremalose.

2.4 Nitrogen Compounds

Nitrogen compounds play a crucial role in winemaking, affecting various aspects of wine quality and stability. Total nitrogen in must and wine includes one inorganic form and many organic forms. In the next section of the book, we will see analytically how we calculate the total nitrogen content in wines and musts. The content of total nitrogen in musts and wines varies from year to year, apparently for reasons due to cultivation care and climate. An additional factor that affects the nitrogen content in must and wine is the variety of grapes, their origin and the winemaking method and technique we are going to follow, for example, in red wines we have almost twice the total nitrogen content compared to that of white wines. The content of the various nitrogenous compounds in the wines varies between 500 and 4000 mg/L. This quantity constitutes approximately 20% of the dry residue. To calculate pure N_2 nitrogen, we must know that 1 g of N_2 corresponds to 6.25 g of nitrogenous substance. Knowing this we can come to the conclusion that the pure nitrogen N_2 contained in the wines varies between 80 and 640 mg/L.

2.4.1 Inorganic Nitrogen Compounds

The type and content of inorganic nitrogenous compounds in the fruit and consequently in must and subsequently in the wine is constantly changing. Its basic form is in the form of ammonia salts. For the science of oenology, the content of fresh must in ammonium cations (NH_4^+) is very important because it is immediately assimilable by the yeasts. The initiation and completion of alcoholic fermentation largely depends on it. Too little nitrogen can lead to slow or stuck fermentation, while excessive nitrogen (above 450–500 mg/L) can result in undesirable compounds and wine spoilage. In this kind of musts, the addition of diammonium phosphate $(NH_4)_2HPO_4$ or diammonium sulfate $(NH_4)_2SO_4$ is recommended to ensure the start of alcoholic fermentation, its smooth progress and its completion.

2.4.2 Organic Nitrogen Compounds

Organic forms of nitrogen are derived from natural sources such as plant residues, animal manures, sewage, and soil organic matter. Organic form of nitrogen in must and wines is found in various forms. These forms are:

2.4.2.1 Proteins

In general proteins are large biomolecules composed of one or more long chains of amino acid residues. They play a crucial role in various biological processes within organisms. In wines, the type of proteins found belongs to the category of glycoproteins. Glycoproteins are proteins which contain oligosaccharide chains covalently attached to amino acid side chains. Protein levels in grapes have been found to be between 100 and 800 mg/L [9]. The effects of proteins on the organoleptic characteristics of wines are as follows:

1. They enhance the volume and structure of the wine
2. They help to create aromatic compounds
3. They create good quality foam in sparkling wines.

The concentration of proteins in wines depends on the following factors:

1. Grape variety
2. Climatic conditions
3. Cultivation care
4. Winemaking method

Proteins are of particular technological interest in the science of oenology, as they are responsible for the formation of turbidity observed especially in white wines that are relatively poor in tannins. The protein haze is due to the colloidal nature of the proteins, which causes them to precipitate under the influence of heat or the presence of tannins. To avoid protein haze in our wines, we should do checks on their protein stability. How this is done will be explained in a later chapter. It is important to know that the protein stability of a wine is affected by the total amount of proteins, their nature, the active acidity (pH), and quantity of phenolics in our wine [10].

2.4.2.2 Amino Acids

Amino acids play a crucial role in wine, both in terms of flavor and quality. From a chemical point of view, amino acids are the compounds of the formula NH_2–R–COOH. Their molecular weights are below 200. Amino acids are the building blocks of proteins, and they naturally occur in grapes. During wine fermentation, yeast utilizes these amino acids as a source of nitrogen for growth. Amino acid content in wine is influenced by yeast strain, fermentation treatments, grape variety, and terroir. Amino acids make up about 40% of the total nitrogen content in wines. New research studies showing they have a strong influence on the sensory properties of red wine building sweetness and viscosity, and increasing the flavor intensity associated with full-bodied wines. To get a better idea about the qualitative and quantitative detection of amino acids in wines, we can study the table [11] (Table 2.2).

The most important amino acids found in grape juices and wines, as has been reported by several authors, are proline (Pro), arginine (Arg), glutamine (Gln), glutamic acid (Glu), and alanine (Ala) and their concentrations ranged from 10 to 2400 mg/L [21].

2.4.2.3 Peptides (Oligopeptides and Polypeptides)

Peptides are short chains of amino acids linked by peptide bonds [22]. Chains of fewer than 20 amino acids are called oligopeptides, and include dipeptides, tripeptides, and tetrapeptides. Polypeptides have molecular weights under 10,000 Da.

Table 2.2 Amino acids in wines with their concentrations from various researches

Amino acid	White wine (mg/L)	Red wine (mg/L)
Alanine	2.10–238 [12, 13]	0.10–90.2 [14–16]
Arginine	3.55–1075 [12, 13]	0.00–153 [14–17]
Asparagine	0.66–86.7 [12, 15, 18]	1.86–75.8 [14–17]
Aspartic acid	3.9–74.8 [12]	0.28–80.9 [14, 15, 19]
Citrulline	0.00–32.6 [15]	2.23–26.5 [14, 15]
Cysteine	0.13–6.73 [13, 15]	0.94–8.93 [5, 15, 17, 19]
γ-Aminobutyric acid	1.46–444 [12]	1.33–61.44 [15, 16, 19]
Glutamic acid	4.9–140.2 [12, 20]	5.69–112 [14–16]
Glycine	2.1–42.00 [12, 18, 20]	1.30–45.3 [14–17]
Histidine	1.02–79.4 [12]	0.06–75.2 [14–16]
Isoleucine	0.00–18.0 [12]	0.14–10.89 [14, 17]
Leucine	1.40–44.5 [12, 13]	0.20–16.22 [14, 16]
Lysine	1.91–78.8 [12, 13, 15, 18]	1.15–46.9 [14–17]
Methionine	0.22–37.3 [12, 13, 15]	0.00–33.2 [14, 15, 17]
Omithine	0.00–55.0 [12, 15]	0.43–67.2 [15, 16]
Phenylalanine	0.92–52.5 [12, 13]	0.40–29.71 [14, 16, 17, 19]
Proline	24.9–1391 [15, 20]	17.86–4415 [14–17, 19]
Serine	1.14–48.0 [12, 15]	0.18–31.9 [14–17]
Threonine	1.16–62.6 [12, 13]	0.75–19.3 [14–16]
Tryptophan	0.00–9.84 [12, 18]	0.54–12.29 [14, 17]
Tyrosine	0.68–45.9 [12, 13, 15]	0.26–9.18 [14, 16, 17]
Valine	0.00–37.5 [12]	0.11–18.2 [14, 15, 17]

Polypeptides over 10,000 Da are called Proteins. This specific category of chemical compounds constitutes 60–80% of the total nitrogen in wines.

2.4.2.4 Amide Nitrogen

This chemical compounds are characterized by the amide group and have a general formula R-$CONH_2$. The most typical types of amides are asparagine, urea, glutamine as well as ethyl carbamate which it is produced industrially by heating urea and ethyl alcohol and it's a known carcinogen chemical compound [23].

2.4.2.5 Nucleic Nitrogen

The presence of this chemical compounds in wines includes nucleosides, nucleotides, and nucleic acids. Nucleotides, especially 5′-nucleotides, are significant flavor enhancers present in various foods and drinks. However, their specific impact on the flavor of lees-aged wines has not been thoroughly investigated. Only ribonucleotides were found in yeast autolysate and Champagne wines. The highest concentration of total nucleotides was quite low, reaching a maximum of around 3 mg/L in wines aged on yeast lees for 9 years.

2.4.2.6 Bioamines

Bioamines with formula R-NH_2 are organic nitrogen compounds that naturally occur during winemaking. These amines are formed from the decomposition of proteins by micro-organisms. In wines, the most abundant amines are:

- Histamine which is associated with allergic reactions and can cause headaches, flushing, and nasal congestion. Usually, we found it in red wines and aged wines.
- Putrescine contributes to wine bouquet and produced during fermentation. It's present in both red and white wines.
- Tyramine has a bitter taste and is produced during malolactic fermentation by some bacteria.
- Ethanolamine is less studied but can also be traced in wines.

Concentrations of these amines in wine are generally below 10 mg/L, so their levels are typically within safe limits for consumption [24].

2.4.2.7 Amino Sugar Nitrogen

An amino sugar is a type of sugar molecule in which a hydroxyl group –OH has been replaced with an amine group –NH_2. Very small quantities of glucosamine and galactosamine have been traced in protein form of nitrogen.

2.4.2.8 Pyrazines

Pyrazine is a heterocyclic aromatic organic compound with the chemical formula $C_4H_4N_2$. They called methoxypyrazines. We found them in grape skin and stems. Pyrazines are responsible for many of the "green" flavors in wine. These flavors range from fresh herbs to vegetal and bell pepper-like aromas. Some grapes varieties like Sauvignon Blanc, Cabernet Franc, Cabernet Sauvignon, Merlot, and Malbec have higher proportions of Pyrazines. This is no accident since these grape varieties are genetically related. In these varieties, with proper management of the foliage in the vineyard, we can also influence the aromatic profile of the wines.

Pyrazines have an important role in wine's overall flavor profile [25]. It can be negative or positive. In negative, aspects of Pyrazines' presence can be smell like old asparagus, or steamed green pepper, or a well-known wine fault called bell pepper aroma. In positive, aspects of same chemical compound presence can be fresh herbaceous profile like chocolate mint or sweet basil.

2.5 Phenolic Compounds

Phenolic compounds are the compounds that contain in their molecule the characteristic group of phenol. These compounds play an important role in the science of oenology. They are the compounds responsible for the differences between red and white wines. They are of great technological interest due to their bactericidal and antioxidant action and many modern studies and research have proven that phenolic compounds are vital for the human body due to their anti-aging and anti-inflammatory properties. The antioxidant activity of phenolic compounds results from the elimination of free radicals by donating a hydrogen atom to radicals. The antioxidant and antiradical activity of these compounds is positively correlated with the number of hydroxyl groups attached to the aromatic ring. The use of sulfate salts in winemaking it enhances the antioxidant property of phenolic compounds. Phenolic compounds can be classified into four basic categories.

1. Phenolic acids
2. Flavonoid
3. Anthocyanin's
4. Tannins

2.5.1 Phenolic Acids

Phenolic acids are divided into two categories. Benzoic acids and cinnamic acids (Fig. 2.25 and Table 2.3).

2.5.1.1 Cinnamic Acids

These phenols are the main phenols in grape juice. They are the first compounds that oxidize and technologically create problems in the color of white wines. In white wines, this class of hydroxycinnamic compounds constitutes the main class of

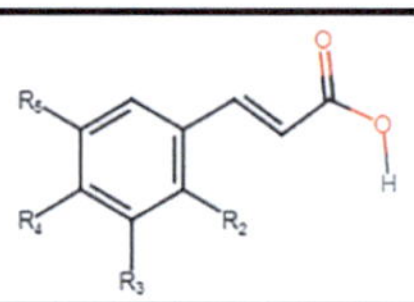

Cinnamic acids (1)

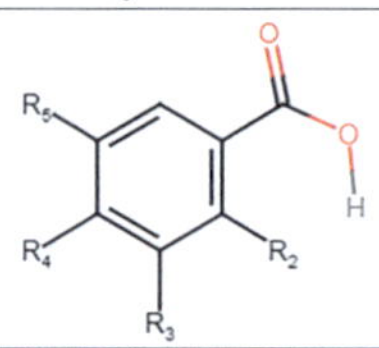

Benzoic acids (2)

Fig. 2.25 Fischer projection of cinnamic acids (1) and benzoic acids (2)

Table 2.3 Benzoic and cinnamonic acids in wines based on R2,R3,R4,R5 formation

Benzoic acids (2)	R2	R3	R4	R5	Cinnamonic acids (1)
p-hydroxybenzoic acid	H	H	OH	H	*p*-coumaric acid
Protocatechuic acid	H	OH	OH	H	Caffeic acid
Vanillic acid	H	OCH_3	OH	H	Ferulic acid
Gallic acid	H	OH	OH	OH	–
Syringic acid	H	OCH_3	OH	OCH_3	–
Salicylic acid	OH	H	H	H	–
Gentisic acid	OH	H	H	OH	–

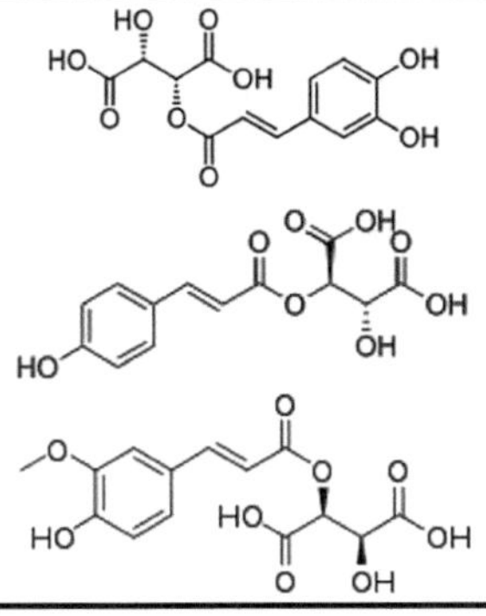

Fig. 2.26 Fischer projection of caftaric acid (1), cutaric acid (2), fertaric acid (3)

phenolic compounds. In this category, there are three compounds found in wine and grapes, *P*-coumaric acid, caffeic acid, and ferulic acid. It must be emphasized that in grapes, the simple hydroxycinnamic acids as previously mentioned do not exist in this form. These acids are found as esters of tartaric acid. In oenology, we have adopted trivial names for these compounds such as caftaric acid, cutaric acid, and fertaric acid, respectively (Fig. 2.26).

These compounds are found in all wines but their levels vary. Caftaric acid is the most prevalent of the group with a concentration in wine grapes of about 170 mg/kg. Cutaric acid in wine grapes is found at a concentration of approximately 20 mg/kg. Finally, fertaric acid is found in wine grapes in concentrations of about 5 mg/kg. Levels of total hydroxycinnamic compounds in the final wine are usually 130 mg/L in white wines and 60 mg/L in red wines [26].

2.5.1.2 Benzoic Acids

Benzoic acids are not found in free form in wine grapes, they exist in the form of complex chemical compounds. Anthocyanins also participate in these complex chemical compounds. Complex chemical reactions contribute to the appearance of free benzoic acids, the aging time being an important factor. Their levels in red wine average close to 70 mg/L, while white wines average close to 10 mg/L.

2.5.2 Flavonoids

The term flavonoids describes a wide group of natural products that form a carbon skeleton with 15 carbon atoms arranged in two aromatic rings joined by a three-carbon bridge (C_6–C_3–C_6). Flavonoids are divided into individual groups such as anthocyanins, flavones, flavonones, dihydroflavonols, chalcones, flavonols, flavans, proanthocyanidins, and isoflavonoids. Flavonols are essentially the yellow pigments of plants. Depending on the form of the lateral ring, we find three forms, kaempferol, quercetin, and myricetin (Fig. 2.27).

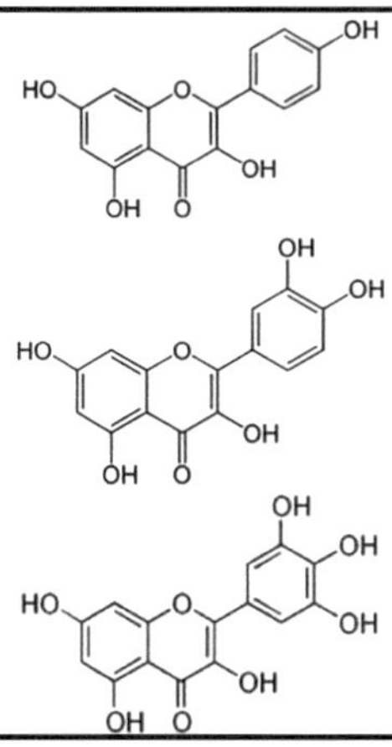

kaempferol (R=R'=H) (1)

quercetin (R=OH, R'=H) (2)

myricetin (R=R'=OH) (3)

Fig. 2.27 Fischer projection of kaempferol (1), quercetin (2), myricetin (3)

We must emphasize the fact that these components are found in grapes in the form of either monoglycosides-3 or monoglucuronosides-3. These are formed in grapes when a monosaccharide molecule (mainly glucose) or a glucuronic acid molecule attaches to position 3 of the central ring of the three molecules above. When a monosaccharide molecule is bonded, the monoglycosides kaempferol-3-monoglucoside, quercetol-3-monoglucoside, and myricetol-3-monoglucoside are, respectively, obtained. In the case of bonding the glucuronic acid molecule, a monoglucuronoside called quercetol-3-mono-glucuronoside results. In red wines where during vinification the stems are in contact with the must during fermentation for a long time, the monoglycosides contained in the grapes are easily hydrolyzed and thus free molecules of kaempferol, quercetin, and myricetin are present.

These three flavonols are detected in traces in the white wines that during their production, the grapes do not come into contact with their juice for a long time. This leads us to the conclusion that flavones, despite the fact that they are the yellow pigments of plants, do not contribute to the yellowish shades of white wines.

2.5.3 Anthocyans

Anthocyanins are essentially the red pigments of grapes. They focus mainly on the bark but we also find them in the flesh of some varieties. They are also found in large quantities in the leaves of the vines especially toward the end of the growing season. From a chemical point of view, they are substances that are derivatives of phenyl-2-benzopyryl. This molecule contains, as we can see in the image, a pyryl molecule in the form of a cation which allows its reaction with other anions and consequently the formation of salts. This is how the various anthocyanidins of the grape arise which, depending on the form of the compounds in the R3′ and R5′ position, are distinguished into delphinidin, cyanidin, malvidin, pelargonidin, peonidin, and petunidin (Table 2.4).

Table 2.4 Fischer projection of selected anthocyanidins and their substitutions

Anthocyanidin	R^3 & $R^{4'}$ = –OH	$R^{3'}$	$R^{5'}$	R^5	R^6	R^7
Delphinidin	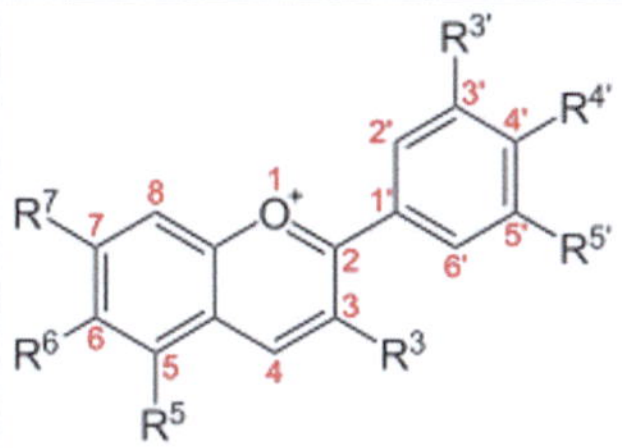	OH	OH	OH	H	OH
Cyanidin		OH	H	OH	H	OH
Malvidin		OCH_3	OCH_3	OH	H	OH
Pelargonidin		H	H	OH	H	OH
Peonidin		OCH_3	H	OH	H	OH
Petunidin		OH	OCH_3	OH	H	OH

The anthocyanidins detected in the grape. The natural pigments present in the grape are chemically not anthocyanid molecules but anthocyanid compounds with glucose molecules attached to the R3 position of anthocyanidin or simultaneously to the R3 and R5 positions. When this happens, monoglycosides and diglycosides result. The new molecules resulting from the union of the anthocyanid molecule with glucose molecules are called anthocyanins or anthocyanins. Anthocyanins are much more stable compounds than the precursor anthocyanidins. At this point, we must emphasize the fact that we have monoglycosides in the case of European varieties (*Vitis vinifera*) (Table 2.5).

In the cases where diglycosides result, i.e., a compound of glucose molecules in the R3 and R5 position of the respective anthocyanidin molecule, we have the compounds of Table 2.6. We must emphasize the fact that we have diglycosides in the cases of wines resulting from crosses of *Vitis vinifera* and of the species *Vitis riparia* and *Vitis rupestris*. This fact leads us to the conclusion that depending on the type of anthocyanins detected in the wines, a distinction can be made between wines originating from European grape varieties of the *Vitis vinifera* species and wines originating from the American grape varieties of the Vitis species riparia and *Vitis rupestris*. Recent studies have detected traces of diglycosides in European grape varieties of the species *Vitis vinifera*, so we conclude that quality control of anthocyanins is not in itself a measure of discrimination between European grape varieties and American varieties. To reach a safe conclusion of discrimination, we should take into account the quantification of diglycosides [26].

The shade of anthocyanins is affected by the pH, the addition of sulfuric anhydride, the molecular structure, and the environmental conditions (Reducing or non-reducing).

More specifically, the more acidic the wine, the more intense red hue it has. When the pH increases, then the anthocyanidins acquire a colorless form and the redness decreases. This reaction is reversed, i.e., if we artificially lower the pH value again, the red color returns. With the addition of sulfuric anhydride, colorless

Table 2.5 Fischer projection of Monoglycosides resulting from the union of anthocyanides with a glucose molecule at the R^3 position

Anthocyanins	$R^{4'} = -OH$	R^3	$R^{3'}$	$R^{5'}$	R^5	R^6	R^7
Delphinidin monoglycoside-3	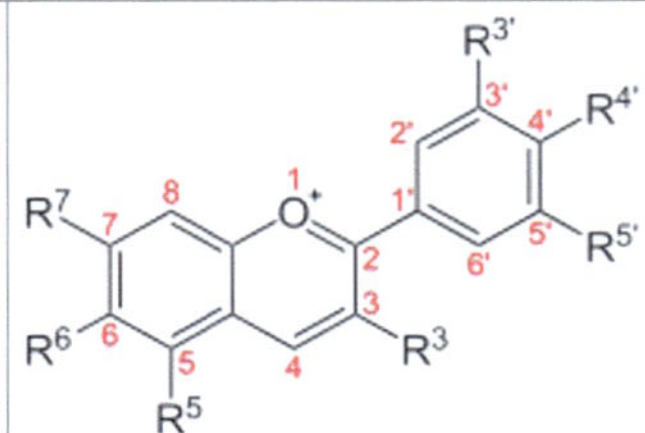	$O–C_6H_{11}O_5$	OH	OH	OH	H	OH
Cyanidin monoglycoside-3		$O–C_6H_{11}O_5$	OH	H	OH	H	OH
Malvidin monoglycoside-3		$O–C_6H_{11}O_5$	OCH_3	OCH_3	OH	H	OH
Pelargonidin monoglucoside-3		$O–C_6H_{11}O_5$	H	H	OH	H	OH
Peonidine monoglucoside-3		$O–C_6H_{11}O_5$	OCH_3	H	OH	H	OH
Petunidine monoglucoside-3		$O–C_6H_{11}O_5$	OH	OCH_3	OH	H	OH

Table 2.6 Diglucosides resulting from the union of anthocyanides with one or two glucose molecules at the R3 and R5 positions

Anthocyanins	$R^{4'}$ = –OH	R^3	$R^{3'}$	$R^{5'}$	R^5	R^6	R^7
Delphinidin di-glucoside-3	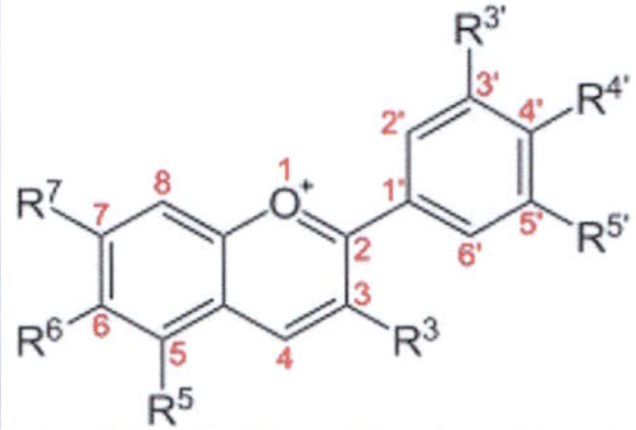	$O–C_6H_{11}O_5$	OH	OH	$O–C_6H_{11}O_5$	H	OH
Di-glucoside-3 cyanidine		$O–C_6H_{11}O_5$	OH	H	$O–C_6H_{11}O_5$	H	OH
Di-glucoside-3 malvidin		$O–C_6H_{11}O_5$	OCH_3	OCH_3	$O–C_6H_{11}O_5$	H	OH
Pelargonidin di-glucoside-3		$O–C_6H_{11}O_5$	H	H	$O–C_6H_{11}O_5$	H	OH
Di-glucoside-3 peonidine		$O–C_6H_{11}O_5$	OCH_3	H	$O–C_6H_{11}O_5$	H	OH
Petunidine di-glucoside-3		$C_6H_{11}O_5$	OH	OCH_3	$C_6H_{11}O_5$	H	OH

compounds are formed with the anthocyanins. This results in partial discoloration of the wines. Due to the gradual reduction of free sulfite, the color of the wines returns. When anthocyanins are in a reducing environment (e.g., Alcoholic fermentation), they become discolored.

From the above we understand that in fresh red and red wines, which at the same time have a low pH, addition of sulfuric anhydride and a reducing environment, their red hue will change considerably over time.

2.5.4 Tannins

From a chemical point of view, tannins are polymerization products of simple phenols. By definition tannins are substances capable of producing stable combinations with proteins and other plant polymers such as polysaccharides. Tannins are molecules with characteristic properties and a large molecular weight ranging from 500 to 3500. In the molecules that have a relatively small molecular weight, there are not enough active sites; as a result, the compounds formed with the protein molecules are unstable. In the opposite case, the molecules of tannins with a high molecular weight cannot get close enough to those of the proteins, resulting in no formation of compounds. Depending on the structure of their molecules, tannins are classified into two categories, hydrolysable and condensed.

2.5.4.1 Hydrolysable Tannins

This category of tannins consists of a glycide on which various phenolic compounds are attached, the main ones being gallic acid and ellagic acid. Hydrolyzed tannins are not contained in the grape but may be found in wines as an oenological additive.

2.5.4.2 Condensed Tannins

This category of tannins is the natural tannins of grapes and wines and come from the polymerization of flavanol-3 and mainly flavanediol-3,4.

2.5.4.3 Properties of Tannins

Because of the great diversity in their chemical structure, tannins have not been adequately categorized. For this reason, they have been categorized according to their properties, which are essentially their ability to join proteins or other polymers, forming insoluble compounds. These properties have important practical applications in various fields, namely:

- They inhibit the action of enzymes, connecting to their protein part.
- They precipitate the proteins of the saliva, as a result of which the feeling of astringency and the bitter taste of the wines is created.
- They participate in the process of clarifying wines in which protein materials are used. More specifically, they unite with proteins forming complexes and due to their weight they settle and drag the various suspensions.

- They have a strong antioxidant effect and protect the wines from the effect of oxygen.
- Tannins polymerize throughout the life of the wines. They form macromolecular complexes which in turn create the pigment precipitates.

2.6 Vitamins

By the term vitamin we refer to chemical compounds that are organic molecules (or a set of closely related molecules called vitamers). These compounds are necessary for the proper functioning of an organism. It is very important to know that these compounds cannot be synthesized in the body in sufficient quantities for survival, this means that somehow they must enter the nutritional cycle of an organism in order for their levels to allow it to function properly [27]. Vitamins are of great technological interest in the science of oenology as they play an important role in the development of the various fermentations carried out during winemaking. Vitamins play a role in various metabolic processes, including those involving key aroma precursors and aromatic compounds. Consequently, they indirectly affect the development of wine aroma profiles by contributing to the synthesis of several important families of flavor molecules. Several studies have been done on their role, but there are still many unanswered questions in various aspects of their metabolic importance. The term vitamin does not include the three other groups of essential nutrients: minerals, essential fatty acids, and essential amino acids [28].

Thirteen compounds or group of compounds are generally characterized as vitamins. According to their physical properties, they are either water-soluble or fat-soluble. Aqueous solvents appear as polar or ionizable groups. Fat-soluble vitamins are more common characterized by aromatic and aliphatic groups [27]. The wine has the peculiarity of coming from a series of successive or simultaneous fermentations. Due to this, there is a great variation in the type and amount of vitamins present in the grapes, then in the grape must and finally in the wine and it is difficult to give specific numbers because different measurements will have to be made at each stage of winemaking. More generally, from the various researches that have been done over the years, above vitamins have a special role in winemaking [29], these are:

- Ascorbic acid (C)
- Thiamine (B1)
- Riboflavin (B2)
- Niacin (B3)
- Pantothetic acid (B5)
- Pyridoxine (B6)
- Biotin (B8)
- Folic acid (B9)
- Cobalamin (B12)
- myo-Inositol

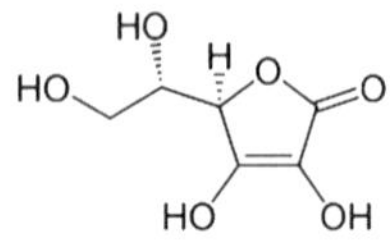

Fig. 2.28 Fischer projection of vitamin C

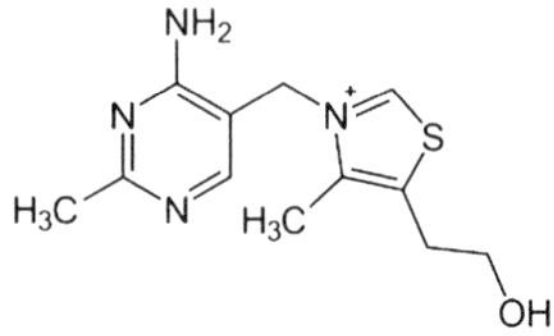

Fig. 2.29 Fischer projection of vitamin B1

2.6.1 Ascorbic Acid (Vitamin C)

Ascorbic acid is an organic compound with formula $C_6H_8O_6$, originally called hexuronic acid. It is a white solid, but impure samples can appear yellowish. It dissolves freely in water to give mildly acidic solutions. It is a mild reducing agent. Vitamin C is the generic descriptor for compounds possessing, qualitatively, the biological activity of ascorbic acid, that is, 2,3-didehydro-L-threo-hexano-1,4-lactone, also occurring in its oxidized form, dehydroascorbic acid [27] (Fig. 2.28).

Ascorbic acid presents a protective effect against must oxidations catalyzed by tyrosinase and laccase, standing as a substrate to the latter [30] limiting their action by monopolizing oxygen through its fast reaction speed, rather than inhibiting the enzymes [31]. The content of ascorbic acid in grape musts has been shown to vary between 30 and 572 mg/L. The initial concentration continuously decreases during the fermentation processes to concentrations ranging between 1 and 30 mg/L [32].

In wines to which ascorbic acid has been added as an oenological additive for antioxidant reasons, the measurement of total and free sulfite is affected and a different method of quantitative determination of sulfite anhydride is required to obtain correct results.

2.6.2 Thiamine (Vitamin B1)

Thiamine, also known as thiamin and vitamin B or aneurin, is a water-soluble vitamin belonging to the B complex, whose trivial designation is 3-(4-amino-2-methyl pyrimidin-5-pyrimidinyl)–5-(2-hydroxyethyl)–4-methylthiazolium [33]. Thiamine is soluble in water, methanol and glycerol, but practically insoluble in less polar organic solvents [34] (Fig. 2.29).

Vitamin B1 is mainly concentrated in the flesh of the grape skin. This has the result that it is contained in higher quantities in red wines [35] than in white wines where the contact of the must with the grapes is of limited duration. It is important to know that vitamin B1 is destroyed by the addition of sulfuric anhydride, by the heating of the must require in some winemaking protocols as well as by the addition

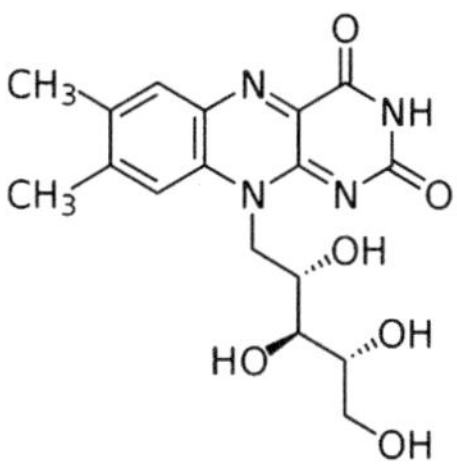

Fig. 2.30 Fischer projection of vitamin B2

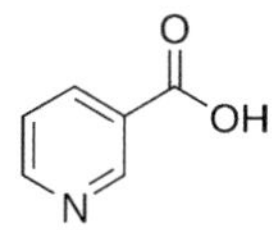

Fig. 2.31 Fischer projection of vitamin B3

of bentonite. Thiamine contents in grape musts have been reported to range between 80 and 1.2 mg/L their average ranging between 0.1 and 1 mg/L.

2.6.3 Riboflavin (Vitamin B2)

Riboflavin is a water-soluble vitamin from the B group, whose trivial designation is 7,8-dimethyl-10-(1′-dribityl)isoalloxazine. Riboflavin is known as vitamin B2, vitamin G, or lactoflavin [27]. Riboflavin is essential to the formation of two major coenzymes, FMN (also known as riboflavin-5′-phosphate) and FAD (flavin adenine dinucleotide) (Fig. 2.30).

Riboflavin contents in grape musts have been shown to range between 3 and 1.45 mg/L, their average ranging between 1 and 100 μg/L, increasing to concentrations between 8 and 133 μg/L in white wines, and 0.47–1.9 μg/L in red wines [36]. It is important to know that vitamin B2 is destroyed during vinification when we add sulfuric anhydride or bentonite. Light also destroys a significant amount of vitamin B2.

2.6.4 Niacin (Vitamin B3)

Niacin, also known as vitamin B3, is the generic descriptor for a B-group water-soluble vitamin and a generic descriptor for pyridine 3-carboxylic acid and derivatives, including both nicotinamide and nicotinic acid, that exhibit a nicotinamide biological activity [33]. Vitamin B3, according to research done in the past, is present in grape must in a content ranging from 0.86 to 2.56 mg/L [37] (Fig. 2.31).

During vinification, the content of vitamin B3 has a special characteristic whereby a decrease is observed in the first stages of the beginning of the alcoholic fermentation and at the end of the alcoholic fermentation a return of its content close to its initial concentration is observed. This is due to the fact that this vitamin

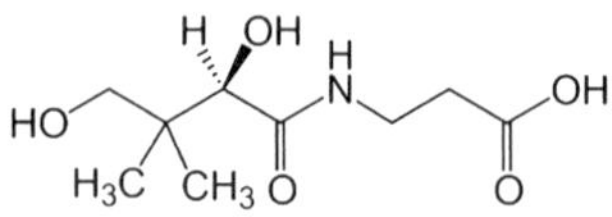

Fig. 2.32 Fischer projection of vitamin B5

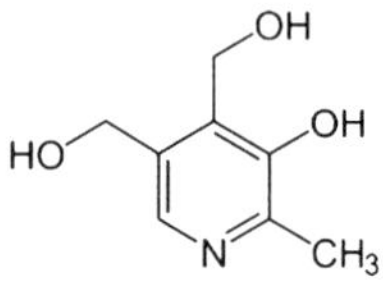

Fig. 2.33 Fischer projection of vitamin B6

is used by the yeasts in the initial stages of fermentation, but after the end of it and due to the autolysis of the yeast cells, they return to the wine.

2.6.5 Pantothetic Acid (Vitamin B5)

Pantothenic acid, also named vitamin B5, is a water-soluble vitamin, whose trivial designation is dihydroxyβ,β-dimethylbutyryl-β-alanine, formerly named pantoylβ-alanine [38]. Grapes were reported to contain an average of 8.5 and 6.8 mg/L pantothenic acid in red and white cultivars [39] (Fig. 2.32).

The special feature of vitamin 5 in relation to winemaking is that wine has a higher content of this vitamin than grape must, this is because vitamin B5 is synthesized by yeasts.

2.6.6 Pyridoxine (Vitamin B6)

Pyridoxine, known as vitamin B6, is one of the eight water-soluble vitamins in the B complex. It is characterized by a tetra-substituted pyrimidine ring, which is connected to a methyl group, a hydroxyl group, and two methyl-hydroxyl groups [40]. The vitamin exists in two active forms: pyridoxamine-5′-phosphate and pyridoxal-5′-phosphate (PLP), both of which can be interconverted [33] (Fig. 2.33).

The concentration of pyridoxine in grapes averages 1.25 mg/L for red varieties, whereas it varies and decreases to 0.88 mg/L for white varieties [41].

2.6.7 Biotin (Vitamin B8)

Biotin, also referred to as vitamin B8, is a water-soluble vitamin, commonly known as cis-hexahydro-2-oxo-1H-thieno[3,4-d]imidazole-4-pentanoic acid. Its structure consists of an imidazole ring fused to a sulfur-containing tetrahydrothiophene ring, which is further substituted with a valeric acid chain [42]. The term "biotin" originates from the German word "Biotin" and is derived from the Ancient Greek word "βίοτος" (bíotos), meaning "life," combined with the suffix "-in," which is

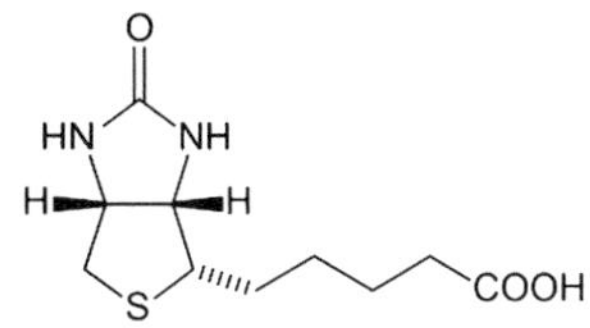

Fig. 2.34 Fischer projection of vitamin B8

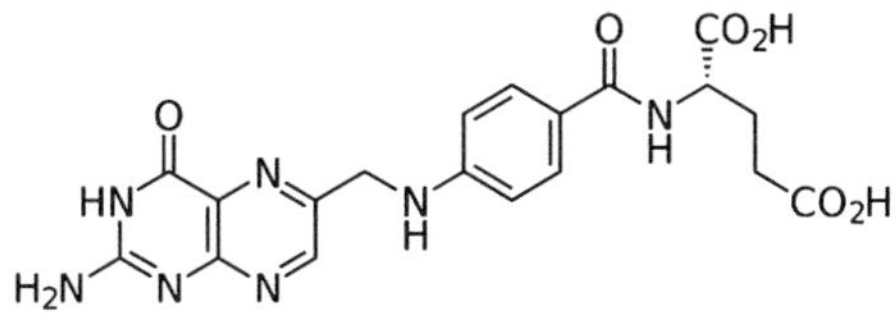

Fig. 2.35 Fischer projection of vitamin B9

commonly used in chemistry to denote "forming." [43] Biotin appears as a white, needle-like crystalline solid (Fig. 2.34).

It has been reported that grapes contain lower amounts of biotin compared to other fruits. Concentrations of biotin vary between vine cultivars. Red grapes have been found to contain higher levels of biotin compared to white grapes. In red grape must, the biotin content indeed averages 2.85 μg/L, while in white grape must it decreases to 1.47 μg/L [44].

2.6.8 Folic Acid (Vitamin B9)

Folic acid, also known as pteroylmonoglutamic acid or vitamin B9, is a term encompassing all compounds with similar biological activity. This group of heterocyclic compounds is derived from the *N*-[(6-pteridinyl)methyl]-*p*-aminobenzoic acid structure, combined with glutamic acid residues [33] (Fig. 2.35).

The folic acid content in grape musts varies from 3 to 50 μg/L, with most reported values being several micrograms per liter. These concentrations slightly decrease in wines, reaching levels from 0.4 to 4.5 μg/L [45].

2.6.9 Cobalamin (B12)

Vitamin B12, also known as cobalamin, is a water-soluble vitamin. Vitamin B12 stands as the most chemically intricate of all vitamins. The structure of B12 is based on a corrin ring and the central metal ion is cobalt. Vitamin B12 in wines varies between 0.05 and 0.1 μg/L (Fig. 2.36).

2.6.10 myo-Inositol

Myo-Inositol, the predominant form of inositol, is a water-soluble compound structured as a hydroxylated cyclic six-carbon ring, commonly known as

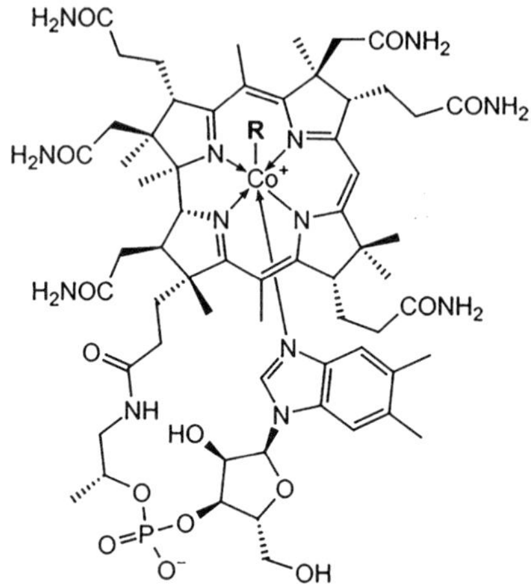

Fig. 2.36 Fischer projection of vitamin B12

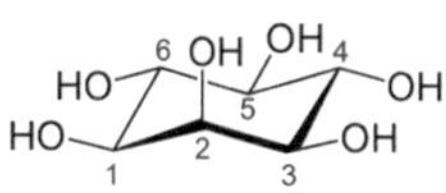

Fig. 2.37 Fischer projection of myo-Inositol

cis-1,2,3,5-trans-4,6-cyclohexanehexol [33]. Inositol levels in grape musts have been reported to vary from 340 to 710 mg/L, which is adequate to fully meet the requirements of yeast (Fig. 2.37).

2.7 Enzymes

Numerous enzymes are found in must and wine. By the term enzymes are meant the chemical compounds which act as catalysts of the biological systems and speed up the various reactions. In oenology, as in life and nature in general, enzymes have a prominent place. All enzymes of direct interest to us in our science are proteins. For more details about them, we can refer to the corresponding chapter on proteins. The enzymes present in must and wine either come from the grapes or produced by various micro-organisms.

In this specific chapter, we will deal with those enzymes which we consider more important. But let's take things in order. From the moment the grape is cut from the plant and more specifically from the moment the grape breaks and begins the process of turning it into wine, a series of biochemical and chemical transformations of its original components begin.

Invertase is the enzyme that hydrolyzes the disaccharide sucrose into the fermentable monosaccharides glucose and fructose. Invertase is found in abundance in grape skins and is also produced as a by-product by yeasts. The action of invertase for our science is decisive.

Oxidases, or oxygen-transferases, are enzymes that come from the grape and catalyze the transfer of oxygen to the various phenols of the must that make up the substrate. In grapes, there are two oxidases of great technological interest. These are tyrosinase and laccase. Tyrosinase is found in all grapes whether they are in good or bad quality. In particular, it is attached to the chloroplasts of the grape cells. Its enzymatic activity focuses on o-diphenols and o-monophenols [46].

Laccase is present in grapes that are rotten. The amount of laccase is completely independent of the degree to which the grapes have been affected by fungal diseases. Laccase exhibits the same enzymatic activity as tyrosynase and additionally acts on *p*-phenols. These two enzymes are responsible for the consumption of oxygen in the whey. In order to remove the largest amount of these two enzymes, it is recommended to de-strain the wine at the beginning of the winemaking process. This is because the largest amounts of tyrosynase and laccase are found in the first sludges.

Pectinases are enzymes that degrade pectin, a polysaccharide located in plant cell walls, via hydrolysis, transelimination, and deesterification reactions. Pectinases are essential in the winemaking process for clarification and extraction. These enzymes decompose pectin, a complex polysaccharide present in the skins and cell walls of grapes, which aids in juice extraction and enhances clarity.

Glycosidases are a class of enzymes which catalyze the hydrolysis of glycosidic bonds in complex sugars. Glycosidases play a crucial role in liberating aromatic compounds, like terpenes, from the glycosides found in grapes. These compounds significantly enhance the aroma and flavor of wine.

Proteases are enzymes that catalyze proteolysis, the breakdown of proteins into smaller polypeptides or single amino acids, which in turn leads to the formation of new protein products. These enzymes aid in protein degradation. By dismantling proteins, they help prevent haze formation and enhance the stability of wine.

2.8 Inorganic Components

Researches that has been done on wines regarding the amount and type of inorganic components contained in wines, has shown that they contain approximately 2–4 g/L of them. The presence of inorganic elements in wine is influenced by numerous factors such as soil type, climate conditions, fruit maturity, pesticide usage, extraction processes, storage, and technological factors. The analysis of inorganic anions in food samples is important from the nutritional, toxicological and technological point of view [47].

These inorganic components are formed in the vine and specifically in their stems and end up in the wine. For a better presentation, we have separated these compounds into cations and anions.

2.8.1 Anions of Wines

2.8.1.1 Chloride Cl^-

Chloride is among the most common inorganic anions found in foods. Consequently, its measurement in food products is essential to comply with legal standards and quality control criteria. The concentration of chloride ions in wine can differ significantly between varieties, yet it seldom surpasses 500 mg/L. These variations are indicative of the local soil and water conditions, with proximity to the coast being a significant influence. Sodium chloride, commonly known as table salt, plays a role in shaping the taste and quality of wines. Dry Extract and Minerals. Factors Affecting Chloride Levels are the geographic, geological, and climatic conditions of vine cultivation affect the levels of chloride and sodium ions in wines. It's interesting that wines from vineyards near the sea or those using tap water with high chlorine content tend to have elevated chloride levels. Chloride may be utilized alongside egg albumin-type natural proteins as a flocculating agent for wine clarification or for organoleptic adjustments to enhance flavor. Moreover, some vintners in dry regions employ traditional practices such as barrel washing with seawater or saline solutions to raise the chloride content in wine [48].

2.8.1.2 Sulfite SO_4^-

In the wine there is a small amount of sulfates that comes from the vineyard, this concentration varies from 0.1 to 0.4 g/L of K_2SO_4. The presence of sulfate in wines depends on the use of sulfiting agents. These are necessary to prevent oxidation during winemaking and to inhibit bacterial growth. However, excessive use of sulfiting agents can impair the wine's flavor. The artificial addition of H_2SO_4 to wines results in a noticeable increase in sulfate content [48]. At this point it is useful to mention that the use of sulfite in foods is being questioned because of its allergenicity, which may cause asthma, dermatitis, urticaria, bronchoconstriction, or anaphylaxis insensitive humans [49]. Legislation regarding the addition of sulfuric anhydride is strict and must be followed closely. In general, there is a new trend in wines that wants consumers to prefer wines whose sulfite content is as low as possible.

2.8.1.3 Phosphorus PO_4^-

Phosphorus is one of the most important inorganic anions in wine, as it participates in various metabolic and chemical reactions. Phosphates in wines typically correlate with the levels in their musts. The phosphate content can be augmented by adding ammonium phosphate, ammonia glycerophosphate, and calcium phosphate to the must, these are sanctioned additives used as yeast activators. Generally, red wines exhibit higher phosphate levels than white wines. It is reported that the phosphate level in red wines ranges from 75 to 1500 mg/L depending on the region, while the level in white wines is usually lower than in reds [50]. In some cases, during winemaking and specifically at the beginning of it, we add ammonium phosphate to the must in order to facilitate the alcoholic fermentation. At this point, we must know that in white wines, the addition of Ammonium phosphate combined with the presence of air can form ferric iron. So we have to be quite careful because possible formation of trivalent iron can lead to future cloudiness. To avoid such a case, we must either do an iron analysis to know precisely its concentration, or even more simply, instead of using Ammonium phosphate, add ammonium sulfate to the wine must.

2.8.1.4 Fluoride F^-

Compounds containing fluoride, such as the natural mineral cryolite (sodium aluminum fluoride), have been utilized for years in controlling pests in vineyards. The timing and rate of application are crucial in reducing fluoride levels on the fruit. Concentrations of fluoride nearing 3 mg/L can potentially cause issues during and after fermentation. No significant differences have been recorded between the fluoride content of white and red wines, but the reported difference exists depending on the place of origin of the wine.

2.8.1.5 Silicon (Si)

Silicon, in the form of silicic acid, is present in wines at concentrations of 30–70 mg/L. It fortifies the surface tissue of the vine and enters the wine through the berry skins, where it exists in a biologically active, assimilable form. Wines with higher silicon content are known to have a sedative effect, such as the rosé wines from the Irouléguy region of France, located near the Spanish border in the southwest.

2.8.1.6 Boron (B)

Boron is a crucial element for vine nutrition. Boric acid may be found in wine in amounts ranging from 10 to 120 mg/L, but most of the boron is precipitated with tartrate crystal compounds as deposits, thereby being removed from the wine itself. For the grapevine, however, boron plays a vital role in photosynthesis and carbohydrate supply. It is also actively involved in pollination, flowering, and fertilization processes.

2.8.2 Cations of Wines (Minerals)

2.8.2.1 Potassium

Potassium is the most significant cation in wine, with concentrations varying from 0.1 to 1.8 g/L. Potassium acid tartrate is the predominant form found in wines. This form of potassium is technologically significant due to its impact on the organoleptic properties of the wine, such as the sour and salty taste, as well as the clarity of the wine through the formation of crystalline sediment.

Apart from its natural occurrence as potassium acid tartrate, potassium metabisulfite ($K_2S_2O_5$) is a crucial additive for wine preservation. The quantity of Potassium Metabisulfite added for proper wine preservation varies, depending on the winemaking type, from 0.2 to 0.4 g/L. Human intervention in augmenting the potassium content of wines, through additives, is noteworthy due to potassium's significant technological interest. Additionally, potassium is applied as a fertilizer in the form of potassium sulfate, which supports the normal functioning of leaves and the root system.

2.8.2.2 Sodium

Sodium typically occurs in concentrations of 20–200 mg/L in wine. Its levels may rise due to the addition of SO_2 as sodium-containing salts or through the use of low-quality bentonite. Moreover, vineyards situated near the sea often produce wines with higher sodium content. In some cases oenologists often use a technique known as sodium ion exchange to clarify the wine and reduce sediment. This process introduces a small amount of salt into the wine.

2.8.2.3 Calcium

The calcium content in grapes varies annually and differs from one vineyard to another. The composition of the soil, including its limestone content, structure, and hydraulic properties, significantly influences the amount of calcium available and its transportation through the vine into the fruit and wine. The calcium content in wine typically ranges from 50 to 100 mg/L. The concentration of calcium in wine is always lower than in musts due to its removal as sediment. Neutral calcium tartrate precipitates more slowly than acidic potassium tartrate, leading to most clarity issues in bottled wines being attributed to neutral calcium tartrate. Calcium is involved in the precipitation of colloids formed between phosphoric acid and ferric iron, and in the tannins and gelatin used during the fining of wines. Additionally, its concentration may increase in wines because $CaCO_3$ is used to reduce high acidity. Calcium nitrate, added as a fertilizer, promotes the synthesis of sugars, fruiting, aromatic substances, and root system development.

Wines with calcium concentrations exceeding 60 mg/L for red wines and 80 mg/L for white wines are more prone to calcium instability. There are primarily three methods to extract calcium from wine. The first and most efficient is through ion exchange, which removes calcium ions but necessitates ion exchange equipment and could affect the wine's sensory properties. The second method involves blending the wine with another that has lower calcium levels. The final method is to cool the wine and introduce it to high-quality, finely ground calcium L-tartrate crystals, a process dependent on the availability of these crystals.

2.8.2.4 Magnesium

Wine appears to have a mean magnesium content of about 114 mg/L. Magnesium is added as a fertilizer in the form of magnesium sulfate, since it participates in the formation of chlorophyll and activates many enzymes involved in fruiting and the metabolism of sugars [51]. In general, wines contain more magnesium compared to calcium, as magnesium is not lost during fermentation, besides its salts are soluble and their level is maintained at the same level during the life of the wine.

2.8.2.5 Iron

Iron is a primary metal in wines, and its impact on wine clarity has been extensively studied. Along with copper (Cu), iron is responsible for major clarity issues in wines. Other metals like aluminum (Al), tin (Sn), and nickel (Ni) can cause minor problems. Clarity issues arise when a wine's iron content exceeds 15–20 mg/L, leading to turbidity or sediment. These issues manifest when the wine is oxygenated and contains high levels of tannin, resulting in blue turbidity, or phosphoric acid, causing white turbidity. Iron exists in wine in various forms.

In an anaerobic environment, iron exists in a divalent state, whereas in wines that have been oxygenated for any reason, the divalent iron is gradually transformed into trivalent iron. Divalent iron compounds are always soluble in wine and do not affect its clarity. However, trivalent iron compounds may be soluble or insoluble. It is the insoluble ferric iron compounds that cause clarity issues in wine. Specifically, ferric iron in wine can form various compounds that fall into one of the following categories [52]:

- Either converted to $Fe(OH)_3$.
- Whether it participates in the formation of soluble compounds with the hydroxy-acids citric, malic, etc. without creating clarity problems.
- Combines with phosphoric acid to form insoluble salts of phosphoric acid ($FePO_4$) (White haze).
- It unites with the phenolic compounds of the wines—mainly with the anthocyanins—and forms blue or black colored compounds. (Cyan or black blur).

Iron in the form of chelated iron, a water-soluble form of iron, is sometimes added as a fertilizer. It is essential for the biosynthesis of chlorophyll.

2.8.2.6 Copper

Copper's role in wine has been extensively researched due to its significant technological implications. In wines, when the copper concentration exceeds 0.5 mg/L in a reducing environment, monovalent copper dominates. This form of copper reacts with sulfuric acid, leading to haziness. It's crucial to note that the balance between monovalent and divalent copper shifts depending on the presence or absence of oxygen. Monovalent copper, which causes copper haze, is predominant under reducing conditions. However, if a wine with this issue is exposed to an oxygen-rich environment or treated with H_2O_2, the existing haze will gradually clear. Furthermore, the development of copper haze is hastened by the wine's exposure to diffused sunlight, indicating a photochemical process is at play.

Copper has a strong presence in crops in general since it is used as a fungicide but apart from that itself as an element is an ingredient useful enzyme. It plays a role in the composition and stability of chlorophyll, participates in the metabolism of carbohydrates as well as proteins [52].

2.8.2.7 Aluminum

Aluminum, according to research studies that have been carried out, has a content of about 10 mg/L in the wines. Aluminum is in increased doses in wines and they come from clay soils. Aluminum in the form of aluminum sulfate is sometimes used as a fertilizer in cases where there is a deficiency in the soils where the vineyard is established. White wines generally show higher aluminum content compared to red wines.

2.8.2.8 Zinc

According to individual research on the content of various heavy metals in wines, the content in wines ranges from 350 to 1850 μg/L. The large variations have an explanation since we usually have high levels of zinc in case of bad use of zinc sulfate for clarifying in combination with potassium ferricyanide [53]. Also zinc in the form of zinc sulfate is used as a fertilizer because it directly corrects zinc deficiency.

2.8.2.9 Manganese

The manganese content, according to researches that have been done, varies in red wines from 0.435 to 7.836, in white wines from 0.674 to 2.203 and in rosé wines from 0.844 to 1.805 [54]. Manganese is found in high concentrations in the pips of the grape, but because these have a high resistance to pressure by their nature, we can hardly say that in the wine we have an increase in the concentration of manganese which comes from their oppression. Manganese, in the form of manganese sulfate, is added as a fertilizer. It is a chemical element that participates in the synthesis of chlorophyll, in nitrogen metabolism and makes plants more productive.

2.8.2.10 Arsenic

According to individual research on the content of various heavy metals in wines, the content in wines ranges from 0.59 to 3.7 μg/L. Arsenic is found in increased content in wines from vineyards where arsenic derivatives have been used for plant protection purposes. In particular, some producers use arsenic trioxide (As_2O_3) as a pesticide, insecticide, or herbicide.

2.8.2.11 Lead

According to individual research on the content of various heavy metals in wines, the content in wines ranges from 1.7 to 18.0 μg/L. In the literature on the subject to date, lead is mainly treated as a result of human industrial activity. Until the beginning of the decade 1990, Lead was a component of fuel. In addition, industrial air pollution was an additional reason for the concentration of Lead in the vineyards [55]. Studies have concluded that Brass (a lead alloy widely used in traditional wine cellars) was the main source of lead contamination in wines.

Modernization of the wineries and the replacement of brass with stainless steel had the effect of reducing the content of lead in the production of wine. However, wines produced today still have significant levels of Lead. It is very important to know the source of its origin, in order to be able to improve this important issue with corrective actions. However, it is also very important to find a material that will not affect technologically the quality of the wines and will provide a solution to the issue of heavy metals in general [56].

Lead is elevated in wines from vineyards where lead derivatives have been used for plant protection purposes. It is also suspected that the use of lead capsules (which has now been banned) may be one of the reasons why there are increased amounts of lead in some wines [53].

2.8.2.12 Cadmium (Cd)

According to individual research on the content of various heavy metals in wines, the content in wines ranges from 0.16 to 0.42 μg/L. The cadmium content in wine is influenced by various factors, both natural and exogenous. Natural factors include the soil type and composition where the vineyards are situated, the grape variety, and the climate. Exogenous factors refer to those that come from external sources during the winemaking process. Elevated concentrations in wine samples are often attributed to the use of pesticides or fertilizers containing cadmium salts. Many studies have documented the treatment of soil and the spraying of vineyards and fruit with cadmium-containing substances [57].

2.9 Aromatic Components

Although the aromatic characteristics of wines consist of chemical compounds that could fit into previous categories in the book, we have chosen to create this separate chapter for them because wine aromas are a huge asset to the science of oenology. Virtually all the cultivation care that takes place in the vineyards and the winemaking process aims to create wines with fine organoleptic characteristics. Wine aromas can be perceived by nose or in the mouth via postnasal way, and is a direct function of the chemical composition of the wine. The perceived flavor of wine is the outcome of intricate interactions among all volatile and non-volatile compounds it contains [58]. The aroma of wine is extremely complex and composed of hundreds of chemical compounds [59].

Wine flavors can be categorized into the following classes:

- Varietal aroma
- Pre-fermentative aroma
- Fermentative aroma
- Post-fermentative aroma

2.9.1 Varietal Aroma

Varietal aroma is the characteristic of the grape variety. The primary compounds contributing to wine's aroma, and largely its characteristic flavor, are the varietal or primary flavors derived from the grapes. These, along with the phenolic compounds' secondary metabolites from the vineyard, play a significant role in the wine's profile. Aromatic substances of this category are located in the epicarp, specifically within the cells of the hypodermis and epidermis. Depending on their aromatic potential, *Vitis vinifera* varieties have been classified into two categories:

- Grapes with simple flavor: Wine contains flavoring substances that originate from compounds known as flavor precursors. These compounds are present in certain grape varieties in a bound, odorless form. Examples of varieties that belong to this group include Cabernet Sauvignon, Pinot Noir, Chardonnay, and Sauvignon, which is considered to be a semi-aromatic variety.
- Aromatic grape varieties: Aromatic grape varieties contain chemical odorants that, despite their very low concentrations in the order of ng/L, are perceptible because most are in a free, odorous form. Varieties such as Muscat (Muscat Ottonel, Muscat de Hamburg, Muscat d'Adda, Muscat de Alexandria), Tămâioasa Românească, and Traminer belong to this class [60].

Primary flavoring compounds, or varieties, are present in grape berries both as free aromas and as aroma precursors.

2.9.1.1 Free Aromas

Free aromas are made of olfactory perceptible compounds: terpenic compounds (terpene and terpenoids), methoxypyrazines, and rotundones.

2.9.1.1.1 Terpenic Compounds

Terpenic compounds exhibit tropical and floral flavors and are present in aromatic grape varieties. In grapes, terpenes predominantly occur as monoterpenes, such as α-terpineol, linalool, nerol, geraniol, and citronellol. The terpenic content in wine varies by variety, ranging from very low levels in simple varieties like Feteasca and Cabernet, approximately 2 μg/L, to higher concentrations of around 2000 μg/L in Muscat varieties [61]. Terpenic compounds contribute to a variety of flavors such as rose (geraniol, nerol), coriander, rosewood (linalool), field flowers (linalool oxide), citrus (citronellol), and define the floral flavor characteristic of Muscat (geraniol, nerol, and linalool) [62] (Figs. 2.38, 2.39, 2.40, and 2.41).

2.9.1.1.2 Methoxypyrazines

Methoxypyrazines in the vine is associated with the metabolism of amino acids such as leucine, isoleucine, valine, and glyoxal. Methoxypyrazines are volatile aromas that impart plant-like flavors, reminiscent of green pepper and capsicum. These compounds can negatively affect the quality of wines by imparting an undesirable flavor, or conversely, they can contribute to a complex flavor profile that is characteristic of the Cabernet group and the Sauvignon variety [63] (Fig. 2.42).

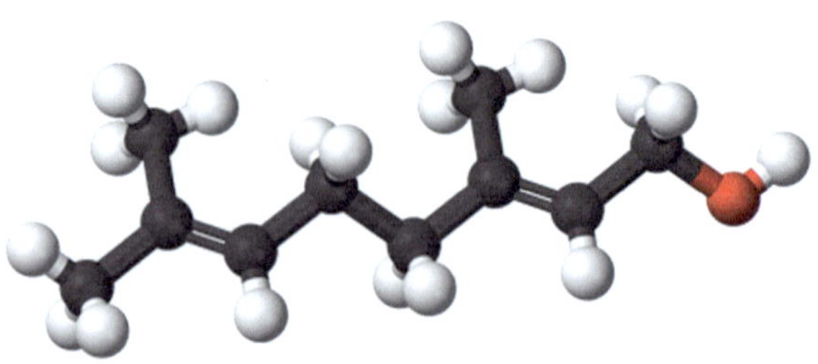

Fig. 2.38 Ball and stick projection of geraniol

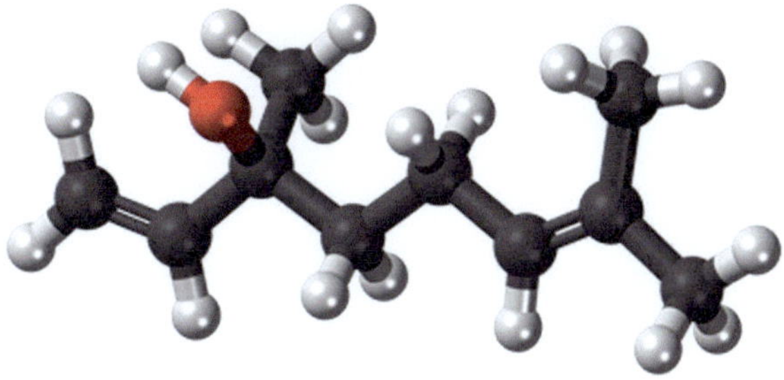

Fig. 2.39 Ball and stick projection of linalool

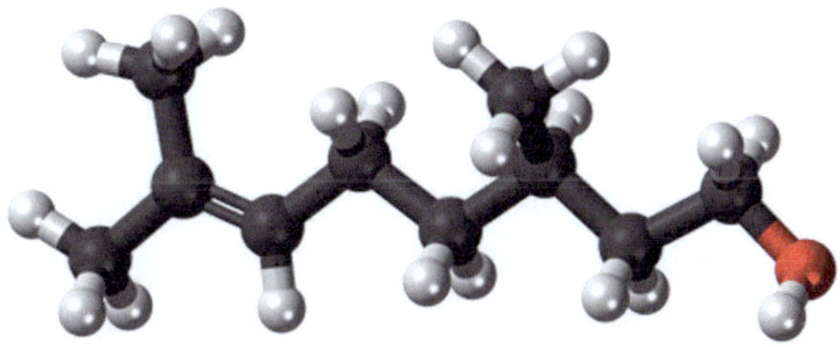

Fig. 2.40 Ball and stick projection of *R*-(+)-Citronellol

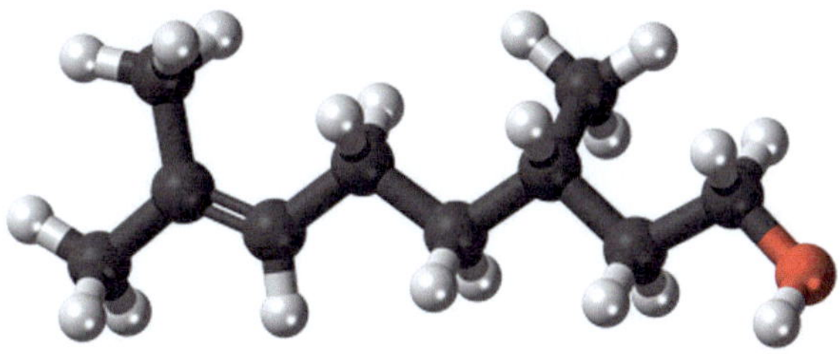

Fig. 2.41 Ball and stick projection of *S*-(−)-Citronellol

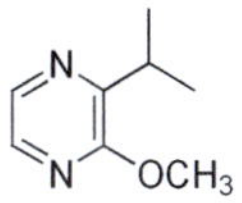

Fig. 2.42 Chemical structure of isopropyl methoxy pyrazine (IPMP)

2.9.1.1.3 Rotundone

Rotundone is a class of free flavor compounds responsible for the spicy notes, such as white and black pepper, that are characteristic of the Syrah grape variety. These highly aromatic molecules have been identified in Syrah wines from Australia, in Italian red wines like Schioppettino, Vespolina, and Groppello di Revò, in Austrian Grüner Veltliner white wines, and in French Gamay and Pineau d'Aunis varieties. Notably, their detection threshold is very low, at 8 ng/L in water and 16 ng/L in wine [64] (Fig. 2.43).

2.9.1.2 Flavor Precursors

Flavor precursors are compounds in a bound form, unnatural and attached to other molecules. These compounds, such as fatty acids, glycosides, and phenolic acids, are cleaved during the processing of grapes and alcoholic fermentation, resulting in

Fig. 2.43 Chemical structure of Rotundone

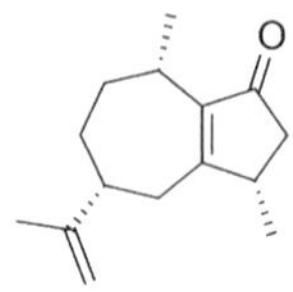

the release of volatile odorants [65]. Aroma precursors are composed of chemically and organoleptically distinct substances, categorized into:

2.9.1.2.1 Glycosidic Precursors

The glycosidic precursors identified in grapes are:

- β-D-glucopyranoside,
- α-L-Ramnosyl:6-O-(α-L-ramnopyranosyl)β-D-glucopyranoside,
- α-L-Arabinofuranosyl:6-O-α-L-arabinofuranosyl)-β-D-glucopyranoside,
- β-D-Apiosil:6-O-(β-D-apofuranosyl)-β-D-glucopyranoside [66].

2.9.1.2.2 Carotenoids

Carotenoids in grapes serve as precursors to C13-norisoprenoids, which exist in free forms as a diverse array of compounds with significant aromatic variety [67]. These are:

- β-damascenone, with flower aroma, exotic fruits, plum jam, apples.
- β-ionone, with violet flavor.
- 1,1,6-Trimethyl-1,2-dihydronaphthalene (TDN) with kerosene flavor.
- Isotypes of vitisperine, with camphor and eucalyptus aroma.
- (E)-1-(2,3,6-trimethylphenyl)buta-1,3-diene with freshly cut grass aroma [68].

2.9.1.2.3 Volatile Thiol Precursors

Thiol components exist solely as non-volatile and odorless precursors. The aromatic character emerges in wine during the fermentation process, as thiol precursors are degraded by *Saccharomyces cerevisiae* yeasts through the action of S-β-lyase enzymes [69]. Among the precursors of volatile thiols identified in grapes, we can mention Cys-4MMP and Cys-3MH, cysteine-linked molecules, and a glutathione-containing 3-mercaptohexanol thiol precursor found in the Sauvignon variety [60].

2.9.1.2.4 Dimethyl Sulfide Precursors

A precursor of dimethyl sulfide [70], a compound responsible for the characteristic aroma of Syrah, is shown to be S-methylthionine, the most likely source of DMS (dimethyl sulfide) in wine [60].

2.9.2 Pre-fermentative Aroma

Pre-fermentative aroma, arising during grape processing. Throughout the processing of grapes and musts, a variety of volatile compounds are produced. These, along with the primary aromas, contribute to the distinctive aroma of young wine, characteristic of the grape variety and the terroir from which it comes.

2.9.3 Fermentative Aroma

Fermentation aroma produced by yeast and bacteria during alcoholic and malolactic fermentations. The yeast *Saccharomyces cerevisiae* acts on terpene compounds through the hydrolysis of glycosides as well as the isomerization and reduction of free terpene alcohols, resulting in the production of aromatic compounds that contribute to the characteristic flavors of wines. Therefore, in addition to the flavors produced during fermentation, other volatile compounds are created through the interaction of the yeast with the primary flavors. Chemicals produced during alcoholic fermentation are higher alcohols, volatile fatty acids, aldehydes, esters, and sulfur compounds. These compounds have a direct or indirect impact on the olfactory characteristics of the wine.

2.9.3.1 Higher Alcohols

Depending on the assimilable nitrogen level for the grape must yeast and the sugar concentration of the grape must, higher alcohols are born through one of two metabolic pathways: -through the Ehrlich pathway, corresponding to the catabolism of the amino acids; -from the anabolic pathway of amino acids, starting from sugars. The concentration of wines in higher alcohols is quite low, 400–500 mg/L, most of them being found in concentrations lower than their sensory threshold of perception.

Higher alcohols traced in wines are, 2-phenylethanol, which has rose floral aroma, ethyl butanoate with aromas of pineapple and strawberries, ethyl hexanoate with aromatic notes of green apples and tropical fruits, ethyl octanoate with floral and fruity odor, ethyl decanoate with fruity and floral essence, ethyl acetate sometimes has negative smell of solvent in wines, butyl acetate with banana aromas, ethyl propanoate with cherries aromas, 2-methylbutyl acetate with fruity aromas, 3-methylbutyl acetate with the classic banana odor, 2-phenylethyl acetate with fruity odor, 2-methylpropanoat of ethyl with aromatic note of red and dark fruits, 2-methyl butanoate of ethyl with aromas of apples and strawberries, 3-methyl butanoate of ethyl with green fruits or blueberries odor, hexyl acetate with exotic fruits smell, linalool acetate with bergamot odor and benzyl acetate with the smell of apple or jasmine (Figs. 2.44 and 2.45) [71].

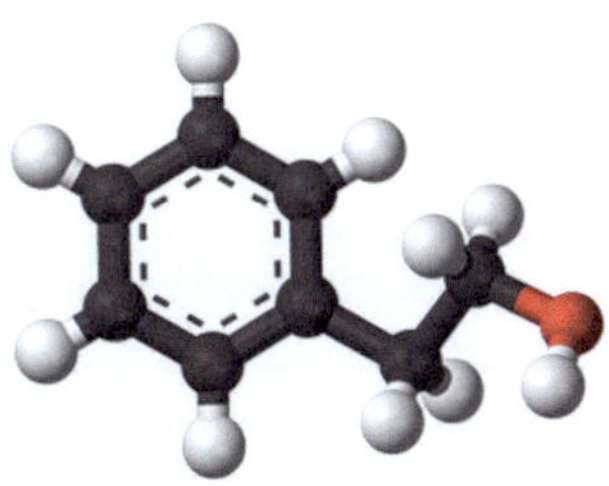

Fig. 2.44 Ball and stick projection of 2-phenylethanol

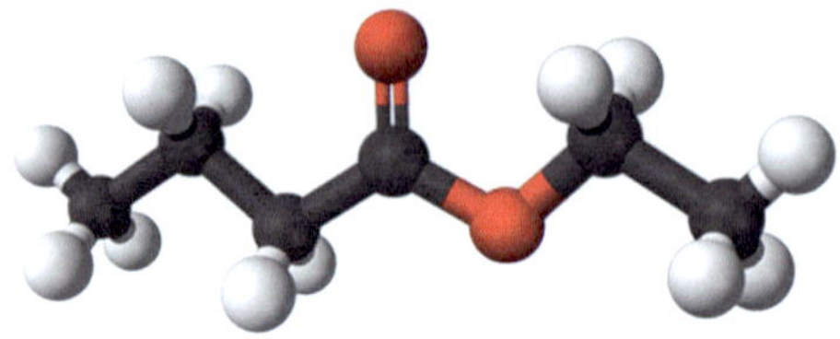

Fig. 2.45 Ball and stick projection of ethyl butanoate

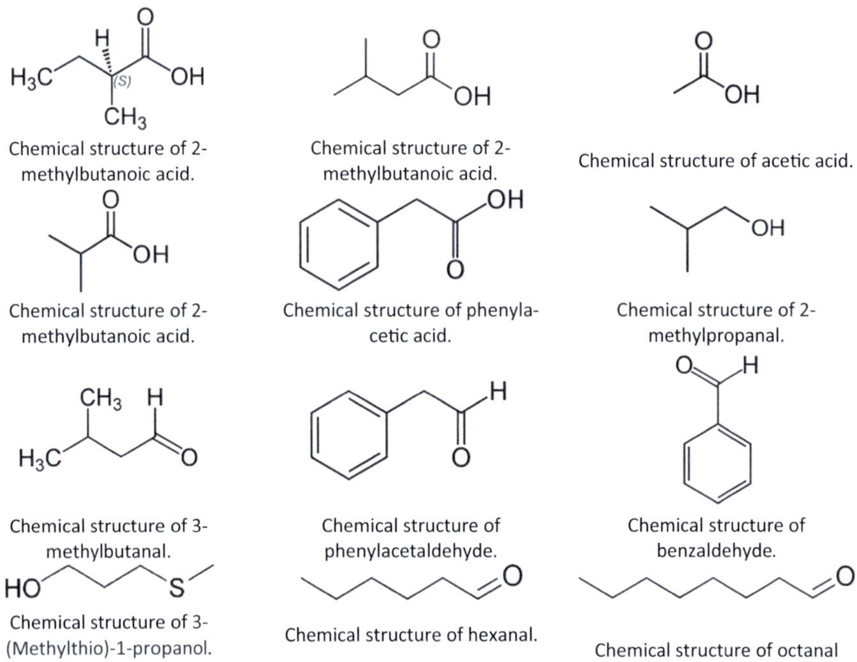

Fig. 2.46 Chemical structures of fatty acids traced in wines

2.9.3.2 Fatty Acid-Aldehydes

The fatty acid group includes short-chain fatty acids produced during yeast alcoholic fermentation through lipid metabolism. Volatile fatty acids are generated from acyl-S-CoA hydrolysis either via the synthetic anabolic pathway or through the β-oxidation of lipids. The production of volatile fatty acids through lipid oxidation takes place at the onset of the fermentation process, influenced by the presence of dissolved oxygen in the must. These fatty acids typically have disagreeable odors; however, their concentrations in wines seldom surpass the sensory threshold, with the notable exceptions of 2-methylbutanoic acid and 3-methylbutanoic acid. Conversely, branched aldehydes, which emit pleasant dried fruit aromas, positively enhance the wine's bouquet (Fig. 2.46).

Fatty acids traced in wines are: acetic acid with the classic smell of vinegar, propionic acid with sour odor, 2-methylpropanoic acid with aromatic note of rot apple, butter or sweat, 2-methylbutanoic acid with fruity aromas, 3-methylbutanoic acid which smell like rotten fruits, phenylacetic acid and hydroxyphenyl acid with floral odor [72].

Aldehydes traced in wines are: Acetaldehyde with ether and sour aroma, 2-methylpropanal, 2-methylbutanal, and 3-methylbutanal with animalic odor, phenylacetaldehyde with rose or floral odor, benzaldehyde with almond smell, 3-(Methylthio)-1-propanol which has a characteristic smell of boiled potatoes, methylthioacetaldehyde with apple odor, hexanal with fruity and froral smell, and octanal with citrus aromatic note [72].

2.9.3.3 Esters

Esters are volatile compounds essential to the aroma of young wines. Most of the esters have floral and fruits odors. During the alcoholic fermentation, three groups of esters are formed: ethyl esters of linear fatty acids are formed through lipid metabolism; ethyl esters of branched or hydroxylated fatty acids; acetates of superior alcohols.

Esters that traced and participate in the aroma of wines are: Ethyl butanoate with fruity odors, ethyl hexanoate with green apple and red fruits aromas, ethyl octanoate ethyl decanoate and ethyl acetate with fruity and floral aromas, butyl acetate with banana flavor, ethyl propionate with cherry odor, 2-methylbutyl acetate with fruity aromas, 3-methylbutyl acetate with fruit and candy odor, 2-phenylethyl acetate with rose smell, 2-methylpropanoat of ethyl with strawberries and blackberries aroma notes, 2-methylbutanoate of ethyl with apple odor, 3-methyl butanoate of ethyl with green and red fruits aromas, hexyl acetate with pear, plum and banana odor, linalyl acetate with bergamot smell and benzyl acetate with jasmine, apple, and pear aromas (Fig. 2.47) [73].

2.9.3.4 Sulfur

The sulfur compounds present in the wine come from both grapes and from the fermentation process through sulfur and nitrogen metabolism [74]. Among sulfur

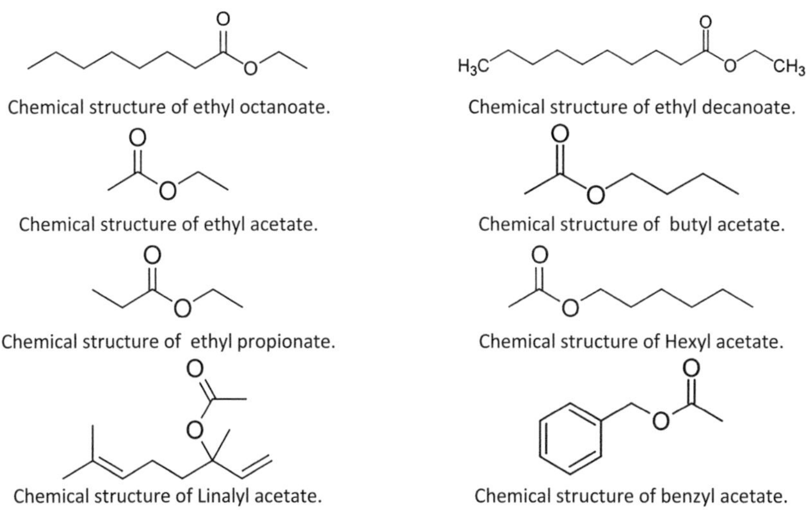

Fig. 2.47 Chemical structures of esters traced in wines

compounds, mercaptans (volatile thiols) and sulfur fermentation compounds are significant.

Mercaptans are a critical group of sulfur compounds that contribute to the aroma of wines, often accompanied by exotic fruit scents. The concentration of sulfur fermentation compounds in wine usually does not surpass their olfactory perception threshold; their contribution to wine's fermentation aroma depends on their molecular weight and volatility.

Sulfur compounds can be categorized, based on these characteristics, into light and heavy or higher compounds. The "lightweight" compounds are linked to the olfactory flaws in wines, producing an unpleasant, off-putting aroma (such as "reduced" smell, rotten eggs, mushrooms, garlic, etc.). In contrast, "heavy" or "superior" compounds like 2-methylthioethanol, 3-methylthiopropanol, and their derivatives (corresponding thioethers and disulfides) contribute a complex aroma to wine. These compounds are actively involved in the aroma of young wines, with concentrations around 0.1 mg/L.

The sulfur fermentation compounds traced in wines are: Thioethyl acetate with sulfurous aromas, thiomethyl acetate with cheese and rotten plants odor, acetyl 2-thiazoline smell like popcorn, methyl-3-propionic acid with grill smell and flavor of maderization benzothiazole has a characteristic rubber smell, dimethyl disulfide smell like asparagus, ethan thiol has the onion aroma, furan methane-2-thiol has coffee odor, hydrogen sulfide smell like rotten egg, 2-mercapto-ethanol has the burned rubber smell, methyl thio-2-ethanol smell like cauliflower, methyl-2-tetrahidrotiofenone smell like gas, and dimethyl sulfide smell like asparagus [60].

2.9.4 Post-fermentative Aroma

Post-fermentative aroma, resulting from changes that take place during the wine's conservation and aging process. In matured wines, the flavor is enhanced by compounds that develop during the aging process, creating what is known as the wine bouquet [75].

From the perspective of this olfactory character, the wine aromas were categorized into nine distinct flavor series. These are as described from Table 2.7.

Table 2.7 Categories of wine aromas based on flavor series

Series	Flavor type
Animal	Smell of meat, musk
Balsamic	Resin, amber, conifer
Chemical	Sulfur, phenol, mercaptan
Wood	Oak, bark, cedar
Empireumatic	Smoke, coffee, cocoa, leather
Spicy	Pepper, mint, cinnamon, anise
Floral	Violet, peony, rose
Fruity	Blueberries, cherries, pears, plums, strawberries, pineapples, bananas
Plant	Grass, hay, olives, tobacco

References

1. Ribereau-gayon P, Dubourdieu D, Donèche B, Lonvaud A (2000) Handbook of Enology, volume 1: the microbiology of wine and vinifications. Wiley
2. Yoshikawa H et al (2013) Investing in our future: The evidence base on preschool education. Foundation for Child Development and Society for Research in Child Development. http://fcdus.org/sites/default/files/Evidence%20Base%20on%20Preschool%20Education%20FINAL.pdf
3. Sundaramoorthy R, Hughes AL, El-Mkami H, Norman DG, Ferreira H, Owen-Hughes T (2018) Structure of the chromatin remodelling enzyme Chd1 bound to a ubiquitinylated nucleosome. eLife 7:e35720. https://doi.org/10.7554/eLife.35720
4. Daverey A, Dutta K, Joshi S, Gea T (eds) (2024) Advances in yeast biotechnology for biofuels and sustainability. Elsevier, pp 585–606. https://doi.org/10.1016/B978-0-323-95449-5.20001-5. ISBN: 9780323954495. https://www.sciencedirect.com/science/article/pii/B9780323954495200015
5. Whiting GC (1976) Organic acid metabolism of yeasts during fermentation of alcoholic beverages—a review. J Inst Brew 82(2):84–92. https://doi.org/10.1002/j.2050-0416.1976.tb03731.x
6. Coulter AD, Godden PW, Pretorius IS (2004) Succinic acid—how it is formed, what is its effect on titrateable acidity, and what factors influence its concentration in wine? Aust NZ Wine Ind J 16:16–25
7. Online Etymology Dictionary. Etymonline.com. Archived from the original on 26 Nov 2016. Accessed 25 Nov 2016
8. Miller WA (1857) Part III. Organic chemistry. In: Elements of chemistry: theoretical and practical. John W. Parker and Son, London, pp 52–57
9. Boulton R, Singleton V, Bisson L, Kunkee R (1999) Principles and practices of winemaking. Aspen Publishers, Gaithersberg, MD
10. Siebert KJ (1999) Effects of protein–polyphenol interactions on beverage haze, stabilization, and analysis. J Agric Food Chem 47(2):353–362. https://doi.org/10.1021/jf980703o
11. Gutiérrez-Gamboa G, Garde-Cerdán T, Moreno-Simunovic Y, Pérez-Álvarez EP (2019) 10—Amino acid composition of grape juice and wine: principal factors that determine its content and contribution to the human diet. In: Grumezescu AM, Holban AM (eds) Nutrients in beverages. Academic Press, pp 369–391. https://doi.org/10.1016/B978-0-12-816842-4.00010-1
12. Soufleros EH, Bouloumpasi E, Tsarchopoulos C, Biliaderis CG (2003) Primary amino acid profiles of Greek white wines and their use in classification according to variety, origin and vintage. Food Chem 80(2):261–273. https://doi.org/10.1016/S0308-8146(02)00271-6
13. Bouzas-Cid Y, Díaz-Losada E, Trigo-Córdoba E, Falqué E, Orriols I, Garde-Cerdán T, Mirás-Avalos JM (2017) Effects of irrigation over three years on the amino acid composition of Albariño (*Vitis vinifera* L) musts and wines in two different terroirs. Sci Hortic 227:313–325. https://doi.org/10.1016/j.scienta.2017.05.005
14. Portu J, López R, López-Alfaro I, González-Arenzana L, Santamaría P, GardeCerdán T (2014) Amino acid content in red wines obtained from grapevine nitrogen foliar treatments: consumption during the alcoholic fermentation. Wine Stud 3(1):4475. https://doi.org/10.4081/ws.2014.4475
15. Tuberoso CIG, Congiu F, Serreli G, Mameli S (2015) Determination of dansylated amino acids and biogenic amines in Cannonau and Vermentino wines by HPLCFLD. Food Chem 175:29–35
16. Gutiérrez-Gamboa G, Garde-Cerdán T, Portu J, Moreno-Simunovic Y, MartínezGil AM (2017) Foliar nitrogen application in cabernet sauvignon vines: effects on wine flavonoid and amino acid content. Food Res Int 96:46–53
17. Gutiérrez-Gamboa G, Portu J, López R, Santamaría P, Garde-Cerdán T (2017) Elicitor and nitrogen applications to Garnacha, Graciano, and Tempranillo vines: effect on grape amino acid composition. J Sci Food Agric 98:2341–2349

18. Benito S, Hoffman T, Laier M, Lochbühler B, Schüttler A, Ebert K, Fritsch S, Röcker J, Rauhut D (2015) Effect on quality and composition of Riesling wines fermented by sequential inoculation with non-Saccharomyces and Saccharomyces cerevisiae. Eur Food Res Technol 241:707–717
19. Zhang XK, Lan YB, Zhu BQ, Xiang XF, Duan CQ, Shi Y (2017) Changes in monosaccharides, organic acids and amino acids during cabernet sauvignon wine ageing based on a simultaneous analysis using gas chromatography–mass spectrometry. J Sci Food Agric 98:104–112. https://doi.org/10.1002/jsfa.8444
20. Sánchez-Gómez R, Garde-Cerdán T, Zalacain A, Garcia R, Cabrita MJ, Salinas MR (2016) Vine-shoot waste aqueous extract applied as foliar fertilizer to grapevines: effect on amino acids and fermentative volatile content. Food Chem 197:132–140
21. Bell SJ, Henschke PA (2005) Implications of nitrogen nutrition for grapes, fermentation and wine. Aust J Grape Wine Res 11:242–295
22. Nelson DL, Cox MM (2005) Principles of biochemistry, 4th edn. W. H. Freeman, New York. ISBN 0-7167-4339-6
23. Jäger P, Rentzea CN, Kieczka H (2000) Carbamates and carbamoyl chlorides. In: Ullmann's encyclopedia of industrial chemistry. Wiley-VCH, Weinheim. https://doi.org/10.1002/14356007.a05_051. ISBN 978-3527306732
24. Mir-Cerdà A, Saurina J, Sentellas S (2022) Bioactive amines in wines. The assessment of quality descriptors by flow injection analysis with tandem mass spectrometry. Molecules 27:8690. https://doi.org/10.3390/molecules27248690
25. Ren A, Zhang Y, Bian Y, Liu YJ, Zhang YX, Ren CJ, Zhou Y, Zhang T, Feng XS (2024) Pyrazines in food samples: recent update on occurrence, formation, sampling, pretreatment and analysis methods. Food Chem 430:137086. https://doi.org/10.1016/j.foodchem.2023.137086
26. Ribéreau-Gayon P, Dubordieu D, Donèche B, Glories Y, Maujean A, Lonvaud A (2006) In: Ribereau-Gayon P, Glories Y, Maujean A, Dubourdieu D (eds) Handbook of enology, vol 2. The chemistry of wine stabilization and treatments, 2nd edn. Wiley
27. Combs GF, McClung JP (2017) Chapter 3—General properties of vitamins. In: Combs GF, McClung E (eds) The vitamins (5th edition): fundamental aspects in nutrition and health. Academic Press, pp 33–38. https://doi.org/10.1016/B978-0-12-802965-7.00003-4
28. Maton A, Hopkins J, McLaughlin CW, Johnson S, Warner MQ, LaHart D, Wright JD (1993) Human biology and health. Prentice Hall, Englewood Cliffs. ISBN 978-0-13-981176-0. OCLC 32308337
29. Evers MS, Roullier-Gall C, Morge C, Gobert A, Alexandre H (2021) Vitamins in wine: which, what for, and how much? Compr Rev Food Sci Food Saf 20:2991–3035. https://doi.org/10.1111/1541-4337.12743
30. Dubernet M, Ribereau-Gayon P, Lerner HR, Harel E, Mayer AM (1977) Purification and properties of laccase from Botrytis cinerea. Phytochemistry 16(2):191–193. https://doi.org/10.1016/S0031-9422(00)86783-7
31. Ribéreau-Gayon P, Glories Y, Maujean A, Dubourdieu D (2006) Handbook of enology: the chemistry of wine stabilization and treatments, vol 2, 2nd edn. Wiley. https://doi.org/10.1002/0470010398
32. Genevois L, Ribéreau-Gayon J (1947) Le vin. Hermann & Cie
33. Combs GF, McClung JP (2017) Chapter 11—Thiamin. In: Combs GF, McClung E (eds) The vitamins (5th edition): fundamental aspects in nutrition and health. Academic Press, pp 297–314. https://doi.org/10.1016/B978-0-12-802965-7.00011-3
34. Mahan LK, Escott-Stump S (eds) (2000) Krause's food, nutrition, & diet therapy, 10th edn. W.B. Saunders Company, Philadelphia. ISBN 978-0-7216-7904-4
35. Schanderl H (1950) Die Mikrobiologie des Weines, vol 2. Eugen Ulmer
36. Ribéreau-Gayon J, Peynaud E, Sudraud P, Ribéreau-Gayon P (1975) Sciences et techniques du vin. Dunod Bordus, Paris, p 342
37. Peynaud E, Lafourcade S (1958) Évolution des vitamines B dansle raisin. Qualitas Plantarum et Materiae Vegetabiles 3–4(1):405–414. https://doi.org/10.1007/BF01884069

38. Combs GF, McClung JP (2017) Chapter 16—Pantothenic acid. In: Combs GF, McClung E (eds) The vitamins (5th edition): fundamental aspects in nutrition and health. Academic Press, pp 387–398. https://doi.org/10.1016/B978-0-12-802965-7.00016-2
39. Hagen KM, Keller M, Edwards CG (2008) Survey of biotin, pantothenic acid, and assimilable nitrogen in winegrapes from the Pacific Northwest. Am J Enol Vitic 59(4):432–436. http://www.ajevonline.org/content/59/4/432.abstract
40. Perli T, Wronska AK, Ortiz-Merino RA, Pronk JT, Daran JM (2020) Vitamin requirements and biosynthesis in Saccharomyces cerevisiae. Yeast 37(4):283–304. https://doi.org/10.1002/yea.346
41. Hall AP, Brinner L, Amerine MA, Morgan AF (1956) The B vitamin content of grapes, musts, and wines. J Food Sci 21(3):362–371. https://doi.org/10.1111/j.1365-2621.1956.tb16932.x
42. Perli T, Wronska AK, Ortiz-Merino RA, Pronk JT, Daran JM (2020) Vitamin requirements and biosynthesis in Saccharomyces cerevisiae. Yeast 37(4):283–304. https://doi.org/10.1002/yea.3461
43. biotin | Origin and meaning of biotin by Online Etymology Dictionary. www.etymonline.com. Archived from the original on 22 Aug 2020. Accessed 14 Nov 2020
44. Ough CS, Kunkee RE (1968) Fermentation rates of grape juice: V. Biotin content of juice and its effect on alcoholic fermentation rate. Appl Microbiol 16(4):572–576. http://www.pubmedcentral.nih.gov/articlerender.fcgi?artid=547471%26tool=pmcentrez%26rendertype=abstract
45. Moreno J, Peinado R (2012) Enological chemistry. Academic Press
46. Τσακίρης Α (1998) Οινολογία. Από το σταφύλι στο κρασί. Εκδόσεις Ψύχαλου, Αθήνα. ISBN: 960 792005 8
47. Dugo G, Pellicano TM, Pera LL, Turco VL, Tamborrino A, Clodoveo ML (2007) Food Chem 102(3):599–605. https://doi.org/10.1016/j.foodchem.2006.05.039
48. Dugo G, La Pera L, Pellicano TM, Di Bella G, D'Imperio M (2005) Food Chem 91(2):355–363. https://doi.org/10.1016/j.foodchem.2004.09.001
49. Castro-Marin A, Buglia AG, Riponi C, Chinnici F (2018) Food Sci Technol 93:174–180. https://doi.org/10.1016/j.lwt.2018.03.003
50. Gonzalez HG, De La Torre AH, Leon JJA (1997) Food Chem 60(3):339–345. https://doi.org/10.1016/S0308-8146(96)00340-8
51. Interesse FS, D'Avella G, Alloggio V, Lamparelli F (1985) Mineral contents of some southern Italian wines. II. Determination of Li, Na, Mg, K, Ca, Co, As, Rb, Sr, Ag, Sb, Ba. Z Lebensm Unters Forsch 181(6):470–474
52. Soufleros E (2012) Oenology, science and technology, Thessaloniki
53. Tsakiris A (1998) Oenology from the grape to the wine. Psyhalou Publications
54. Cabrera C, Teissedre P-L, Cabanis MT, Cabanis JC (2000) Manganese determination in grapes and wines from different regions of France. Am J Enol Vitic 51:103–107. https://doi.org/10.5344/ajev.2000.51.2.103
55. Almeida CMR, Vasconcelos MTSD. Lead contamination in Portuguese red wines from the Douro region: from the vineyard to the final product. LAQUIPAI, Departamento de Qui'mica, Faculdade de Cie^ncias, Universidade do Porto, Rua do Campo Alegre, 687, P4169-007 Porto, Portugal
56. Kaufmann A (1998) Lead in wine. Food Addit Contam 15:437–445. [Rosman K, Chisholm W, Jimi S, Candelone J, Boutron C, Teissedre P, Adams F (1998) Leadconcentrations and isotopic signatures in vintages of French wines between 1950 and 1991. Environ Res 78:161–167]
57. Marin C, Ostapczuk P (1992) Lead determination in wine by potentiometric stripping analysis. Fresenius J Anal Chem 343:881–886
58. Fairbairn SC, Smit AY, Jacobson D, Prior BA, Bauer FF (2014) Environmental stress and aroma production during wine fermentation. S Afr J Enol Vitic 35(2):168–177
59. Vilanova M, Genisheva Z, Masa A, Oliveira JM (2010) Correlation between volatile composition and sensory properties in Spanish Albariño wines. Microchem J 95(2):240–246
60. Vişan L, Tamba-Berehoiu R-M, Popa C, Dănăilă-Guidea S, Culea R (2018) Aromatic compounds in wines. Sci Pap Ser Manag Econ Eng Agric Rural Dev 18:423–430

61. Cordonnier R, Bayonove C (1974) Mise en évidence dans la baie de raisin, variété Muscat d'Alexandrie, de monoterpènes liés révélables par une ou plusieurs enzymes de fruits. L'Academie des Sciences, Series D 278:3387–3390
62. Duchêne E, Legras JL, Karst F, Merdinoglu D, Claudel P, Jaegli N, Pelsy F (2009) Variation of linalool and geraniol content within two pairs of aromatic and non-aromatic grapevine clones. Aust J Grape Wine Res 15(2):120–130
63. Darriet P, Lavigne V, Boidron JN, Dubourdieu D (1991) Caractérisation de l'arôme variétal des vins de Sauvignon par couplage CPG-Olfactométrie. J Int Sci Vigne Vin 25(3):167–174
64. Colette N (2010) L'œnologie, 7th edn. Lavoisier
65. Darriet P, (1993) Recherches sur l'arôme et les précurseurs d'arôme du Sauvignon, Thèse de doctorat de l'université de Bordeaux
66. Roland A, Vialare J, Razungles A, Rigou P, Schneider R (2010) Evolution of S-cysteinylated and S-glutathionylated thiol precursors during oxidation of Melon B. and Sauvignon blanc musts. J Agric Food Chem 58(7):4406–4413
67. Bureau S, Razungles A, Raymond L (2000) Effect of vine or bunch shading on the carotenoid composition in Vitis vinifera L. berries. J Sci Food Agric 80(14):2012–2020
68. Flamini R (2010) Grape aroma compounds: terpenes, C13-norisoprenoids, benzene compounds, and 3-alkyl-2-methoxypyrazines, mass spectrometry in grape and wine chemistry. In: Flamini R, Traldi P (eds) Mass spectrometry in grape and wine chemistry. Wiley, Hoboken, pp 97–116
69. Coarer M, Charrier F, Poulard A (1999) Microflore et typicité des vins. ITV France/IFV
70. Deloire A (2007) Métabolismes secondaires: les arômes, Fiche 10, SupAgro Montpellier
71. Nykanen L (1986) Formation and occurrence of flavor compounds in wine and distilled alcoholic beverages. Am J Enol Vitic 37(1):84–96
72. Lonvaud-Funel A, Renauf V, Strehaiano P (2010) Microbiologie du vin: bases fondamentales et applications. Lavoisier
73. Ribereau-Gayon J, Peynaud E, Sudraud P, Ribereau-Gayon P (1972) Traité d'oenologie, sciences et techniques du vin, 1. Analyse et controle des vins. Ed. Dunod, Paris
74. Roland A (2010) Influence des phénomènes d'oxydation lors de l'élaboration des moûts sur la qualité aromatique des vins de Melon B. et de Sauvignon Blanc en Val de Loire, Thèse de Doctorat
75. Hodson G, Wilkes E, Azevedo S, Battaglene T (2017) Methanol in wine. BIO Web Conf 9:02028. https://doi.org/10.1051/bioconf/20170902028

Type of Wines

3

This chapter presents winemaking protocols that we have created and used with very good results. By winemaking protocols, we mean predetermined steps that someone must follow in order to have a quality wine that will be repeated in the following years applying the same protocol [1]. It is quite reasonable and desirable that some may not fully agree with any of the protocols presented. More specifically, to some it may seem excessive to apply such a protocol and to others it may seem that we are not using all the oenological additives we can in order to have the best possible result. Either way, it is taken for granted that by applying any of the protocols presented you will create an exceptionally high-quality wine.

At this point, we should emphasize that in order to implement any of the following winemaking protocols, we should have at least the necessary equipment [2] that includes, Grape crushers and separators, grape press, stainless steel tanks, and tank cooling system. The equipment in question is the minimum required to create an extremely high-quality wine, beyond that there is specialized equipment for special techniques which for the majority of wineries is equipment that their purchase costs cannot be realized. Obviously wineries that have such specialized equipment do not have the need to follow our protocols. We have always believed that the purpose of an enologist is to create the best possible wine he can by utilizing 100% of the equipment available to the winemaker. Obviously they will produce wines of lesser quality [3], but in any case, they are marketed wines and we need to adapt a vinification protocol for them and get the best possible result [4].

We take it for granted that the terminology used in winemaking protocols is familiar, for example, when we read "harvest" [5] in a protocol we all understand that this is the stage at which the producer/winemaker removing the grapes from the vineyard. However, in order to make this decision, apparently a series of cultivation care and analyses have been carried out. In protocols, these are taken for granted, but nothing can be taken for granted in science because this book can be read by a student of oenology who is just learning the basics, so before we start deciphering a

G. Z. Kyzas, F. Papageorgiou, *Oenology in Practice*,
https://doi.org/10.1007/978-3-031-85531-3_3

winemaking protocol, we will try as soon as possible to present a short description as well as some key points of each stage.

3.1 Harvest

Determining the harvest date involves a series of analyses and grape sampling from the vineyard [5]. Key factors to consider include the sugar to acid ratio, the development of mineral and phenolic components, the grape variety, type of wine plan to create, the post-harvest weather forecast, labor availability, and the capacity of vinification tanks [6].

3.2 Transfer Grapes to Winery

The transfer of the grapes from the vineyard to the winery must be done in a short time and if possible the grape must have the lowest possible temperature [7]. The grape, from the moment it is cut from the plant, undergoes various biochemical changes. As is well known, the grapes are harvested either manually or mechanically. In the first case, the harvest must be carried out very early in the morning so that we have the lowest possible temperatures and with as many workers as possible so that it is done as quickly as possible [8]. At this point, it is worth mentioning that the kind treatment of the grapes by the workers in the transport crates is important. It is also important that the transport crates have to be clean and have openings that allow air circulation and that we do not overload them to avoid unwanted crushing of the grapes. Grapes that are spoiled are not legitimate to collect [9].

In the second case of mechanical harvesting, machines ensure a quick harvest, but due to mechanical harvesting, the grapes suffer more and it is very useful to be able to ensure their transport at low temperatures or in an inert atmosphere. This is done by using vehicles that have this technology. If they are not available, mechanical harvesting should also be done in the early hours of the morning.

3.3 Destemming and Crushing Grapes

At this point, the grapes have now been transported to the winery. Ideally, there should be a sorting belt that will help us remove grapes that do not meet the desired winemaking conditions as well as foreign bodies that may have been transported from the vineyard [8]. The grapes via the conveyor belt are led to the grape destemmer that separates the rails from the stems and at the same time breaks the grapes [10].

3.4 Pre-fermenting Cryoextraction

Freezing grapes is a winemaking technique known as cryoextraction that aims to modify the composition of the final wines. This process is done in order to transfer certain compounds from the skin of the grapes to the must due to the unstructured tissues of the grape [10]. The most common cryoextraction techniques are: (i) Ultrafast freezing and (ii) liquid nitrogen freezing [11]. Research studies have shown that wines obtained by liquid nitrogen freezing exhibited more intense aromas due to higher levels of terpenoids, hydroxyl compounds, and fatty acids than both traditional and ultrafast freezing wines. In any case, both freezing techniques produce wines with a more intense aroma compared to those wines obtained by traditional methods [10].

3.5 Pressing Grapes

The purpose of this stage is to press the grapes in order to receive the juice that is on the grape rail. There are several types of presses with membrane presses being the most prevalent, which are actually the most modern. In some cases of vinification, at this stage, the must that comes from the outflow is separated as being of higher quality compared to the must that comes from the pressure [12].

3.6 Mud Removal: Clarification

This stage of vinification is a key point for the evolution of our wine in the future. Our grape has undergone considerable mechanical processing and understandably the must we currently have in our tanks is not completely pure. The purer the must is during fermentation, the better quality wine will result in the end. The composition of the mud is mainly earthy impurities found in the grape, pieces of the grape skin and various polysaccharides and proteins that settle after their union with the tannins and salts [13]. There are different types of desalination that are used, depending on the equipment available to each winery. The types of clarification are static clarification with or without the use of pectinolytic enzymes, clarification with centrifugation, clarification with flotation, clarification with a cyclone, clarification with a vacuum filter, and finally clarification with filtration [14].

3.7 Alcoholic Fermentation

At this stage of vinification, we have pure must (during white and rosé vinification) and must that contains stems (during red vinification). This stage is what gives us the magic of winemaking; essentially in this stage, the yeasts convert the fermentable sugars into ethanol [15].

In all the protocols given below (with the exception of natural wines) at this stage, the must is inoculated with yeasts. By using commercial yeasts, we ensure repeatability in the organoleptic characteristics of the wine, improve the aromas, ensure a smooth alcoholic fermentation, and avoid the predominance of native yeasts that we are not sure of the final result. The must we are going to inoculate already has some native yeasts that compete with the yeasts we choose to add. However, the number of yeasts we add is so large that in the end the strain of yeasts that we wish to carry out the fermentation prevails.

At this stage of winemaking and after analyses that have been done on the must, in addition to yeasts, we add other oenological substances that aim to facilitate the alcoholic fermentation. Depending on the composition of the must, the protocols we propose can be adapted to the needs of each winemaking. For example, a must that has very low total acidity should be corrected before applying each protocol [16].

A very important factor among others of this stage is the temperature at which the alcoholic fermentation is to be carried out. If we wanted a general rule, we would say that the fermentation of white and rosé wines should be done at a temperature of 16–20 °C. For red vinification because we seek greater extraction of components from the grapes, fermentation temperatures between 25 and 30 °C are recommended. Commercial yeasts state the optimum fermentation temperature and should be consulted.

3.8 Fining Wine

At this stage alcoholic fermentation has been completed and we now have a wine that we need to process to stabilize and bring it to the quality we want. We have to make very important decisions such as whether or not we are seeking malolactic fermentation. If we do not want malolactic fermentation, we add sulfite anhydride and we transfer our wine to clean tanks. If we want malolactic fermentation, we leave the wine with the mud and control the evolution of the fermentation. When we find that it is completed, then we separate our wine from the mud into a clean tank and add a sulfur anhydride to protect it [14].

In any case, the removal of wine from muds is necessary to avoid unpleasant odors that may be created. It is important that the wine has to be carried to clean tanks are full to avoid oxidation; if our tank cannot be filled, then we must fill the empty space in an atmosphere of inert gas. Our wine from now on should, after specific analyses described in the next chapter, be stabilized by the use of appropriate oenological additives and either bottled, or lead to barrels, or remain in our tanks for future use [17].

The winemaking protocols proposed below are protocols designed by oenologist Papageorgiou Frangiskos, one of the two authors of the book. These protocols have been applied many times in wineries with excellent results and several of the wines that resulted from the application of these protocols have been distinguished and won many awards.

3.9 Classic Winemaking

3.9.1 Methods and Protocols for Making Classic White Wines

The main characteristic of white vinification is that the alcoholic fermentation takes place in the absence of lees. Also, for better results, the must after pressing must be separated into outflow must and pressure must. During white vinification, there is a great need for protection against oxidations exactly as described in the protocol we have created. In difficult viticulture years in which we have used large amounts of copper and pesticides, it is important to do a laboratory test and similar oenological additions in order to free our must from the large amounts of those that will prevent us from carrying out the alcoholic fermentation and will degrade our quality of the produced wine. It is important in these cases to remove the damaged grapes (Table 3.1).

3.9.2 Methods and Protocols for Making Classic Red Wines

The most basic characteristic of red winemaking is that the alcoholic fermentation, or better yet, the greater part of it, takes place in the presence of grapes. The contact time of the must with the grapes depends on the health and type of grapes and of

Table 3.1 White wines protocol

	Steps	Additions/key points
1	Harvest	– Sampling and analyses – Low temperatures – Choose healthy grapes
2	Transfer grapes to winery	– Quick transportation – Low temperature
3	Destemming and crushing	– Removal of grapes that do not meet the quality criteria
4	Pre-fermenting cryoextraction	– 6 h at 15 °C – Addition of Sulfur dioxide 100 g/tn – Addition of pectolytic enzymes specialized for this stage
5	Pressing grapes	– Separate outflow must from must after pressing in separate tanks
6	Mud removal	– Chemical analyses and corrections of must – Addition of pectolytic enzymes specialized for this stage (pectin esterase, polygalacturonase, and pectate lyase)
7	Alcoholic fermentation	– Addition of yeast nutrients – Addition of yeasts specialized for the grape variety – Addition of tannins selected for their antioxidant properties – Addition of glutathione formulation – Temperature suitable based on yeast specifications – Aeration of the must by stirring the second and third day
8	Fining wine	– Chemical analyses and corrections – Removing clear wine from muds – Stabilizing produced wine (proteins, tartaric, calcium, color) – Removing bitters – Addition of Sulfur dioxide

Table 3.2 Red wines protocol

	Steps	Additions/key points
1	Harvest	– Sampling and analyses – Low temperatures – Choose healthy grapes
2	Transfer grapes to winery	– Quick transportation – Low temperature
3	Destemming and crushing	– Removal of grapes that do not meet the quality criteria – We transfer the must with the grapes to the fermentation tanks – Addition of Sulfur dioxide 100 g/tn – Addition of pectolytic enzymes specialized for this stage (maltodextrine, pectinases, ammonium sulfate)
*	Pre-fermenting cryoextraction	– Only in cases where we have excellent raw material
4	Alcoholic fermentation	– Addition of yeast nutrients – Addition of yeasts specialized for the grape variety – Addition of concentrated tannins proanthocyanidin type – Temperature suitable based on yeast specifications – Aeration of the must by stirring the second and third day
5	Pressing grapes	– Separate outflow must from must after pressing for future blending
6	Completion of alcoholic fermentation	– Addition of yeast nutrients
7	Malolactic fermentation	– We recommend spontaneous malolactic fermentation without the addition of lactic acid bacteria. But if we notice that it is delayed then we add them
8	Mud removal	– Chemical analyses and corrections of must
9	Fining wine	– Chemical analyses and corrections – Addition of Sulfur dioxide – Addition of polysaccharides – Organoleptic corrections – Stabilizing produced wine (proteins, tartaric, calcium, color) – Removing bitters

course on the type of wine we want to create. In any case, this process must be monitored by laboratory tests and the appropriate decisions must be made at the appropriate times. The protocol that we present to you below has been tested many times and its result is amazing (Table 3.2).

3.9.3 Methods and Protocols for Making Classic Rose Wines

Rosé wines come from the varieties that produce red wines. Essentially we follow a protocol very similar to that of white winemaking with red grape varieties. Two vinification protocols are presented below, and their essential difference is that in the first, direct pressing takes place without extraction, while in the second protocol, before the pressing of the grapes, the grapes remain with the must that has been created during their extraction in order to extract the components, and then they are driven in the press (Tables 3.3 and 3.4).

Table 3.3 Light rose wines protocol

	Steps	Additions/key points
1	Harvest	– Sampling and analyses – Low temperatures – Choose healthy grapes
2	Transfer grapes to winery	– Quick transportation – Low temperature
3	Destemming and crushing	– Removal of grapes that do not meet the quality criteria
4	Pre-fermenting cryoextraction	– 6 h at 15 °C – Addition of Sulfur dioxide 100 g/tn – Addition of pectolytic enzymes specialized for this stage
5	Pressing grapes	– Separate outflow must from must after pressing in separate tanks
6	Mud removal	– Chemical analyses and corrections of must – Addition of pectolytic enzymes specialized for this stage (pectin esterase, polygalacturonase, and pectate lyase)
7	Alcoholic fermentation	– Addition of yeast nutrients – Addition of yeasts specialized for the grape variety – Addition of tannins selected for their antioxidant properties – Temperature suitable based on yeast specifications – Aeration of the must by stirring the second and third day
8	Malolactic fermentation	– We recommend spontaneous malolactic fermentation without the addition of lactic acid bacteria. But if we notice that it is delayed then we add them
9	Fining wine	– Chemical analyses and corrections – Removing clear wine from muds – Stabilizing produced wine (proteins, tartaric, calcium, color) – Removing bitters – Addition of Sulfur dioxide

Table 3.4 Rose wines with short extraction protocol

	Steps	Additions/key points
1	Harvest	– Sampling and analyses – Low temperatures – Choose healthy grapes
2	Transfer grapes to winery	– Quick transportation – Low temperature
3	Destemming and crushing	– Removal of grapes that do not meet the quality criteria
4	Pre-fermenting cryoextraction	– 5–24 h at 15 °C – Addition of Sulfur dioxide 100 g/tn – Addition of pectolytic enzymes specialized for this stage – Depending on the intensity of the color we want, we calculate grape variety, maturity, and health condition and extraction time
5	Pressing grapes	– Separate outflow must from must after pressing in separate tanks
6	Mud removal	– Chemical analyses and corrections of must – Addition of pectolytic enzymes specialized for this stage (pectin esterase, polygalacturonase, and pectate lyase)

(continued)

Table 3.4 (continued)

	Steps	Additions/key points
7	Alcoholic fermentation	– Addition of yeast nutrients – Addition of yeasts specialized for the grape variety – Addition of tannins selected for their antioxidant properties – Temperature suitable based on yeast specifications – Aeration of the must by stirring the second and third day
8	Malolactic fermentation	– We recommend spontaneous malolactic fermentation without the addition of lactic acid bacteria. But if we notice that it is delayed then we add them
9	Fining wine	– Chemical analyses and corrections – Removing clear wine from muds – Stabilizing produced wine (proteins, tartaric, calcium, color) – Removing bitters – Addition of Sulfur dioxide

3.9.4 Methods and Protocols for Making Sparkling Wines

For the production of sparkling wines, we can follow two protocols (methods). The first method is the traditional method with the second fermentation take place in the bottle and the second method is the modern method with the second fermentation take place in a closed resistant tank.

For the traditional method, we will present two protocols which have many similarities but one big difference. The second alcoholic fermentation in the bottle is done using special cases containing yeasts. These yeasts, in combination with the natural addition of sucrose, carry out the alcoholic fermentation in the bottle without passing after their autolysis as sediment in the wine. This case has a diameter of 0.65 μm, this has the result that the components of the wine come into contact with the yeasts and we get the desired result, but without the yeasts being able to leave this membrane and thus our wine remains clear. This is very important because the ramuage stage and the removal sediments stage becomes a very simple process, saving a lot of time and better results (Tables 3.5, 3.6 and 3.7).

Table 3.5 High-quality sparkling wines with traditional method protocol

	Steps	Additions/key points
1	Harvest	– Sampling and analyses – Low temperatures – Choose healthy grapes
2	Transfer grapes to winery	– Quick transportation – Low temperature
3	Destemming and crushing	– Removal of grapes that do not meet the quality criteria
4	Pre-fermenting cryoextraction	– 6 h at 15 °C – Addition of sulfur dioxide 100 g/tn – Addition of pectolytic enzymes specialized for this stage
5	Pressing grapes	– Separate outflow must from must after pressing in separate tanks
6	Mud removal	– Chemical analyses and corrections of must – Addition of pectolytic enzymes specialized for this stage (pectin esterase, polygalacturonase, and pectate lyase)
7	Alcoholic fermentation	– Addition of yeast nutrients – Addition of yeasts specialized for the grape variety – Addition of tannins selected for their antioxidant properties – Addition of glutathione formulation – Temperature suitable based on yeast specifications – Aeration of the must by stirring the second and third day
8	Fining wine	– Chemical analyses and corrections
As we can see till now, we are following the typical white wine protocol to make the base wine. (first fermentation) base wines are usually aged for several months before proceeding to the next phase of the process. Traditionally, base wines were aged in large oak barrels. In recent years, however, winemakers have turned to alternative containers like stainless steel tanks, concrete tanks, and even sizable clay jars known as amphorae. Many vintners believe these contemporary vessels allow the wine's fruit flavors to shine through without the influence oak barrels may impart		
Second fermentation preparation for 100 hl of wine with alcohol 9.5–11.5%		
9a	Yeast rehydration 37 °C	– 10 L water – 500 g addition of yeasts for secondary fermentation – 500 g addition of yeast nutrients – Let the mixture rest for 15 min and stir and we repeat 3 times (45 min in total)
9b	Alcohol acclimatation stage 1 Temperature 20 °C	– 5 L water – 5 L base wine – Concentrated must (70 g/L sugar content) till wine density become between 1020 and 1030 g/L – 10 g addition of yeast nutrients – 24 h
9c	Alcohol acclimatation stage 2 Temperature 17 °C	– Adding 26 L of water of 25 °C – 65 L base wine – Concentrated must (50 g/L sugar content) till wine density become between 1020 and 1030 g/L – 25 g addition of yeast nutrients – 36 h

(continued)

Table 3.5 (continued)

	Steps	Additions/key points
9d	Alcohol acclimatation stage 3 Temperature 15 °C	– Adding 45 L of water of 25 °C – 320 L base wine – Concentrated must (30 g/L sugar content) till wine density become between 1010 and 1020 g/L – 25 g addition of yeast nutrients – 36 h
9e	Pre bottling mixture preparation (same day with stage 9f)	– Mix the accimalated mixture with the whole quantity base wine – Add the sugar necessary to develop the desired atmospheres. For pressure of 1 Atm we need 4.25 g of sucrose per liter of wine, so for a classic sparkling wine of 5 Atm we need 20 g of sucrose per liter of wine. We calculate it after mixing base wine with our mixture – Add 20 g/tn of bentonite gel – Add 300 g/tn yeast nutrients
9f	Bottling for second fermentation	– Bottling takes place in special pressure-resistant bottles
10	Remuage (1–4 years)	– The bottle is laid on its side, and as time passes, the sediment gradually settles along the bottle's underside – Slowly rotating and tilting the bottle. (mechanically or manually) – The bottles are then positioned with the neck facing downward, allowing the sediment to settle into the bottle's neck over time
11 Removing sediments Typically, there are two main methods to accomplish this: – The sediment can be removed manually by swiftly uncapping the bottle, which allows the carbon dioxide pressure to expel the sediment, although some wine may be lost in the process – The second technique, known as "à la glace," entails immersing the bottle's neck in a chilled ice-salt bath to freeze the liquid within. Usually, the bottle cap contains a tiny plastic insert, and the mixture of champagne and sediment in the neck solidifies against this insert. Consequently, when the cap is removed, the pressure within the bottle expels the frozen plug of wine and sediment, along with dead yeast cells, resulting in clear wine		
12	Dosage	– We add to the wine a blend of aged champagne with a small amount of cane sugar, citric acid 0.2 g/L and sulfuric anhydride 0.08 g/L. The amount of sugar solution depends on the desired final sugar content * of our product
* Extra brut (0–6 g/L), brut (<15 g/L), extra dry (12–20 g/L), sec (17–35 g/L) demi sec (33–50 g/L), doux (>50 g/L)		
13	Bottling and corking	– After the dosage has been added and the bottle filled to the correct liquid level, the champagne bottle is then ready for corking – We use strong cork to withstand the force from the pressure of carbon dioxide inside the bottle. The cork is softened by heating, and then hammered into the bottle to seal it – Once the cork is inserted, a wire cage known as a "muselet" is secured around it to ensure it stays in position

Table 3.6 High-quality sparkling wines with traditional method using special cases protocol

	Steps	Additions/key points
1	Harvest	– Sampling and analyses – Low temperatures – Choose healthy grapes
2	Transfer grapes to winery	– Quick transportation – Low temperature
3	Destemming and crushing	– Removal of grapes that do not meet the quality criteria
4	Pre-fermenting cryoextraction	– 6 h at 15 °C – Addition of Sulfur dioxide 100 g/tn – Addition of pectolytic enzymes specialized for this stage
5	Pressing grapes	– Separate outflow must from must after pressing in separate tanks
6	Mud removal	– Chemical analyses and corrections of must – Addition of pectolytic enzymes specialized for this stage (pectin esterase, polygalacturonase, and pectate lyase)
7	Alcoholic fermentation	– Addition of yeast nutrients – Addition of yeasts specialized for the grape variety – Addition of tannins selected for their antioxidant properties – Temperature suitable based on yeast specifications – Aeration of the must by stirring the second and third day
8	Fining wine	– Chemical analyses and corrections
As we can see till now, we are following the typical white wine protocol to make the base wine. (First fermentation) Base wines are usually aged for several months before proceeding to the next phase of the process. Traditionally, base wines were aged in large oak barrels. In recent years, however, winemakers have turned to alternative containers like stainless steel tanks, concrete tanks, and even sizable clay jars known as amphorae. Many vintners believe these contemporary vessels allow the wine's fruit flavors to shine through without the influence oak barrels may impart		
9A	Bottling wine	– Bottling takes place in special pressure-resistant bottles
9B	Secondary alcoholic fermentation. (1–4 years)	– 24 g/L of sucrose (sugar) is added to each bottle. From concentrated sucrose solution 500 g/L – Each bottle is inoculated with an initial yeast population of 1.5 × 106 cells/mL – Using special cases containing yeasts – Tapping bottles with metal caps similar to those used on beers
10	Dosage	– We add to the wine a blend of aged champagne with a small amount of cane sugar, citric acid 0.2 g/L and sulfuric anhydride 0.08 g/L. The amount of sugar solution depends on the desired final sugar content * of our product
* Extra brut (0–6 g/L), brut (<15 g/L), extra dry (12–20 g/L), sec (17–35 g/L) demi sec (33–50 g/L), doux (>50 g/L)		
13	Bottling and corking	– After the dosage has been added and the bottle filled to the correct liquid level, the champagne bottle is then ready for corking – We use strong cork to withstand the force from the pressure of carbon dioxide inside the bottle. The cork is softened by heating, and then hammered into the bottle to seal it – Once the cork is inserted, a wire cage known as a "muselet" is secured around it to ensure it stays in position

Table 3.7 Sparkling wines in tanks protocol

	Steps	Additions/key points
1	Harvest	– Sampling and analyses – Low temperatures – Choose healthy grapes
2	Transfer grapes to winery	– Quick transportation – Low temperature
3	Destemming and crushing	– Removal of grapes that do not meet the quality criteria
4	Pre-fermenting cryoextraction	– 6 h at 15 °C – Addition of Sulfur dioxide 100 g/tn – Addition of pectolytic enzymes specialized for this stage
5	Pressing grapes	– Separate outflow must from must after pressing in separate tanks
6	Mud removal	– Chemical analyses and corrections of must – Addition of pectolytic enzymes specialized for this stage (pectin esterase, polygalacturonase, and pectate lyase)
7	Alcoholic fermentation	– Addition of yeast nutrients – Addition of yeasts specialized for the grape variety – Addition of tannins selected for their antioxidant properties – Temperature suitable based on yeast specifications – Aeration of the must by stirring the second and third day
8	Fining wine	– Chemical analyses and corrections
As we can see till now, we are following the typical white wine protocol to make the base wine. (first fermentation) base wines are usually aged for several months before proceeding to the next phase of the process. Traditionally, base wines were aged in large oak barrels. In recent years, however, winemakers have turned to alternative containers like stainless steel tanks, concrete tanks, and even sizable clay jars known as amphorae. Many vintners believe these contemporary vessels allow the wine's fruit flavors to shine through without the influence oak barrels may impart		
9	Second fermentation for 100 hl of wine with alcohol 9.5–11.5%	
9a	Yeast rehydration (100 hl of wine with alcohol 9.5–11.5%)	– 2.5 kg of yeast – 1.5 kg addition of yeast nutrients – 50 L water
9b	Alcohol acclimatation stage 1 25 °C for 6 h	– 50 L rehydrated yeast – 50 L wine – 50 L water – 10 kg sugar – 50 g organic and inorganic (ammonium phosphate) and cell walls
9c	Alcohol acclimatation stage 2 20 °C for 12 h	– 150 L acclimatized mix – 150 L wine – 100 L water – 12 kg sugar – 100 g organic and inorganic (ammonium phosphate) and cell walls

(continued)

Table 3.7 (continued)

	Steps	Additions/key points
9d	Alcohol acclimatation stage 3 20 °C for 24 h	– 400 L acclimatized mix – 400 L wine – 250 g organic and inorganic (ammonium phosphate) and cell walls
9e	Second fermentation 20 °C	– 800 L of acclimatized mix – Base wine – 300 g/tn g organic and inorganic (ammonium phosphate) and cell walls
10	Cooling	– When the pressure reaches 5 Atm cooling tank at −5 °C for some days. – Adding sulfur dioxide 100 g/tn
11	Filtration	– Moving wine to another special durable fermentation tank that withstand high pressures by special filtration – Chemical analyses and corrections
12	Bottling and corking	– We use strong cork and special bottles to withstand the force from the pressure of carbon dioxide inside the bottle – For safety we use wire cage known as a "muselet" is secured around it to ensure it stays in position

3.10 New/Alternative Styles of Winemaking

3.10.1 Methods and Protocols for Making Extra Light Rose Wines

This category of very light rosé wines is a relatively new wine trend. These are very pale rosé wines that are slightly pink in color (like the color of the inner skin of an onion). They are wines that are consumed fresh with relatively high acidity, as their acidity is also the key to maintaining that soft shade that makes them special (Table 3.8).

3.10.2 Methods and Protocols for Making Orange Wines

The simplest way to describe orange wine is that it's produced from white grapes such as Pinot Grigio—known as Ramato in Italy—Ribolla Gialla from Italy, using the red winemaking process. White wines are traditionally made by pressing white grapes soon after harvesting, separating the juice from the skins, seeds, and stems, resulting in a clear liquid that ferments into white wine. In contrast, red wine derives its color from the grape skins. Despite most red grapes having clear juice, similar to white grapes, they are fermented with their skins, allowing the color to infuse the juice.

Hearing the word and seeing the color of an orange wine we think that the only difference is in the color, but the color is not the only characteristic that makes this category of wines stand out. Contact with white grape skins imparts texture from tannins and complexity from other phenolic compounds that affect the taste and feel of the final wine (Table 3.9).

Table 3.8 Extra light rose wines protocol

	Steps	Additions/key points
1	Harvest	– Sampling and analyses – Low temperatures – Choose healthy grapes
2	Transfer grapes to winery	– Quick transportation – Low temperature
3	Destemming and crushing	– Removal of grapes that do not meet the quality criteria
4	Pre-fermenting cryoextraction	– 3 h at 10 °C – Addition of Sulfur dioxide 80 g/tn – Addition of pectolytic enzymes specialized for this stage
5	Pressing grapes	– Separate outflow must from must after pressing in separate tanks
6	Decolorize	– Addition of decolorizing charcoal (300–400 g/tn) – Addition of bentonite (300 g/tn) – 3–5 h with stirring grape must every 30 min in low temperatures (5–7 °C)
7	Mud removal	– Move grape must with filtration to clean tank in low temperatures (5–7 °C) – Chemical analyses and corrections of must (acidity, pH, sugars) – Addition of pectolytic enzymes specialized for this stage (pectin esterase, polygalacturonase, and pectate lyase) – Addition of Sulfur dioxide 40 g/tn
8	Alcoholic fermentation	– Addition of yeast nutrients – Addition of yeasts specialized for the grape variety – Addition of tannins selected for their antioxidant properties – Temperature suitable based on yeast specifications – Aeration of the must by stirring the second and third day
9	Fining wine	– Chemical analyses and corrections – Removing clear wine from muds – Stabilizing produced wine (proteins, tartaric, calcium, color) – Removing bitters – Addition of Sulfur dioxide

Table 3.9 Orange wines protocol

	Steps	Additions/key points
1	Harvest	– Sampling and analyses – Low temperatures – Choose healthy grapes
2	Transfer grapes to winery	– Quick transportation – Low temperature
3	Destemming and crushing	– Removal of grapes that do not meet the quality criteria – We transfer the must with the grapes to the fermentation tanks – Addition of Sulfur dioxide 120 g/tn – Addition of pectolytic enzymes specialized for this stage (maltodextrine, pectinases, ammonium sulfate)

(continued)

Table 3.7 (continued)

	Steps	Additions/key points
*	Pre-fermenting cryoextraction	– Only in cases where we have excellent raw material
4	Alcoholic fermentation	– Addition of yeast nutrients – Addition of yeasts specialized for the grape variety – Addition of concentrated tannins proanthocyanidin type – Temperature suitable based on yeast specifications – Aeration of the must by stirring the second and third day
5	Pressing grapes	– Separate outflow must from must after pressing for future blending
6	Completion of alcoholic fermentation	– Addition of yeast nutrients
7	Malolactic fermentation	– We recommend spontaneous malolactic fermentation without the addition of lactic acid bacteria. But if we notice that it is delayed then we add them
8	Mud removal	– Chemical analyses and corrections of must
9	Fining wine	– Chemical analyses and corrections – Addition of Sulfur dioxide – Addition of polysaccharides – Organoleptic corrections – Stabilizing produced wine (proteins, tartaric, calcium, color) – Removing bitters

3.10.3 Methods and Protocols for Making Non-alcohol Wines

Dealcoholized wine is a type of wine from which all or most of the alcohol has been removed. This can be a minor adjustment for style, such as reducing the alcohol by volume (ABV) from 15 to 14%, or it can be more significant, lowering the ABV to 8% for a low-alcohol product, or even to less than 0.5% for a non-alcoholic variant. The primary goal of dealcoholization is to extract the alcohol after fermentation, resulting in a wine that maintains the taste and characteristics of the original, but with greatly reduced or no-alcohol content. Beverages with low or no alcohol offer fewer calories and serve as a healthier option compared to their alcoholic counterparts, and their popularity is on the rise. Dealcoholized wine is increasingly popular among those who enjoy wine but wish to avoid the adverse effects of alcohol. It's a preferred option for those abstaining from alcohol due to health, religious, or personal reasons. The rise in its popularity can be attributed to various factors and a shift in consumer behavior. First of all today's consumers are more health-conscious and seek alternatives to traditional alcoholic drinks. Dealcoholized wine offers a way to savor the flavor of wine without the potential health risks of alcohol. Then the demand for low and no-alcohol beverages, initially driven by millennials, has now spread across a broader demographic. Dealcoholized wine appeals to many who seek non-traditional alcoholic options for social gatherings, dining out, or quiet evenings at home. It allows people to engage in social rituals without alcohol, reflecting changing social norms.

Recent technological advancements mean that dealcoholized wine now closely mimics the taste of its alcoholic counterpart. Producers are employing methods like dealcoholization to craft non-alcoholic wines that retain authentic flavors. With the growing demand for non-alcoholic options, producers have broadened their range. Consequently, the market now offers a greater variety of non-alcoholic wine choices. Dealcoholized wine has become a popular beverage at social events where traditional wine might not be appropriate due to religious or social considerations. It enables those who refrain from alcohol to partake in celebrations and social functions without the concerns associated with alcohol consumption.

3.10.3.1 How Non-alcohol or Low-Alcohol Wines Are Made

Methods for wine dealcoholization aim to lower or eliminate alcohol content while maintaining the essential flavors and attributes of traditional wine. There are multiple methods available. Winemakers take into account the type of wine when selecting a dealcoholization technique.

3.10.3.2 Vacuum Distillation

In this process, traditional wine is distilled at reduced pressure within a vacuum chamber, which lowers the boiling point of alcohol. As the wine is heated (up to 48 °C), the alcohol evaporates. This vapor is then captured. The remaining liquid is the dealcoholized wine. The Spinning Cone Column (SCC) and packed columns are the two most prevalent and efficient technologies in this field.

3.10.3.3 Spinning Cone Column

The spinning cone method employs a stainless steel column with a central rotating shaft, surrounded by alternating spinning and stationary cones. This technique swiftly and gently extracts the most volatile aroma compounds in the first pass, followed by the removal of alcohol in the second. Afterwards, the aromatic elements are reintegrated into the wine.

3.10.3.4 Packed Columns

Packed columns are composed of two or more vertical stainless steel cylinders, filled with a packing material that provides an extensive surface area for highly efficient distillation. Operating continuously, this method simultaneously extracts aromas and dealcoholizes the wine in one pass. It yields spirit/alcohol, dealcoholized beverage, and essence in a single step, which are then blended to create the final product.

3.10.3.5 Membrane Filtration

Membrane filtration passes wine through a filter with fine pores that trap alcohol molecules, without applying heat. The final stage involves adjusting the filtered wine's composition to fine-tune the flavor balance. A drawback is the gradual reduction of alcohol content. Each pass reduces alcohol by 0.7–1.5% ABV, meaning a 15% wine would require eight passes to reduce the alcohol content to 3%.

3.10.4 Methods and Protocols for Making Natural Wines

By natural wines, we mean wines that at no stage of winemaking do not use oenological additives. It is essentially an arbitrary term that at the moment these series are written there is no legislative framework. Usually such wines are produced by relatively small wineries and come from biological or biodynamic cultivation vineyards. An important part of this wines voltage is that the alcoholic fermentation is done with the indigenous yeasts exists in the grape and are not added by us. This kind of winemaking has a great risk as when something is spontaneous without our intervention then the result is not discounted. Counseling these wines must have high alcohol in combination with high acidity. These two features to coexist must apply appropriate cultivation care to selected grape varieties. Grapes must also be in excellent condition when intended to become natural wines. In the world of wine production, the detainees for this category are divided.

Natural wines have blossoms and sediments that consumers consider a natural consequence, however, some are filtered before bottling and some do not. In our protocols we consider to protect the consumer that wines before bottling must be filtered. The cleanliness of the equipment using appropriate cleaners is generally considered to be a given in winemaking, but the category of natural wines should be particularly emphasis on the intensive cleanliness of all equipment whose must and wine comes into contact (Tables 3.10, 3.11, and 3.12).

Table 3.10 Natural white wines protocol

	Steps	Additions/key points
1	Harvest	– Strict and representative control over the acidity and sugars of the grapes as well as their hygiene – Low-temperatures harvest – Choose healthy grapes – Harvest by hand
2	Transfer grapes to winery	– Quick transportation – Low temperature
3	Destemming and crushing	– Removal of grapes that do not meet the quality criteria
5	Pressing grapes	– Separate outflow must from must after pressing in separate tanks
6	First mud removal	– For a short period of 2–3 h
7	Alcoholic fermentation	– Temperature of fermentation tanks 17–19 °C – Aeration of the must by stirring the second and third day
8	Completion of alcoholic fermentation	– Analyses of sugars till full consumption – Aeration and stirring of the must every day at the end of alcoholic fermentation
9	Malolactic fermentation	– Malolactic fermentation in natural wine is almost sure
10	Second mud removal	– For a short period of 2–3 h
11	Fining wine	– Removing clear wine from muds – Tanks must be completely full – After 2 months of wine stay in the tanks the wine must be bottled for greater protection against oxidation

Table 3.11 Natural red wines protocol

	Steps	Additions/key points
1	Harvest	– Strict and representative control over the acidity and sugars of the grapes as well as their hygiene – Low temperatures – Choose healthy grapes – Harvest by hand
2	Transfer grapes to winery	– Quick transportation – Low temperature
3	Destemming and crushing	– Removal of grapes that do not meet the quality criteria – We transfer the must with the grapes to the fermentation tanks
4	First mud removal	– For a short period of 2–3 h
4	Alcoholic fermentation	– Temperature of fermentation tanks 20–22 °C – Aeration of the must by stirring the second and third day
5	Pressing grapes	– Separate outflow must from must after pressing for future blending
6	Completion of alcoholic fermentation	– Analyses of sugars till full consumption – Aeration and stirring of the must every day at the end of alcoholic fermentation
7	Malolactic fermentation	– Malolactic fermentation in natural wine is almost sure
8	Second mud removal	– For a short period of 2–3 h
9	Fining wine	– Tanks must be completely full – After 2 months of wine stay in the tanks the wine must be bottled for greater protection against oxidation

Table 3.12 Natural rose wines protocol

	Steps	Additions/key points
1	Harvest	– Strict and representative control over the acidity and sugars of the grapes as well as their hygiene – Low temperatures – Choose healthy grapes – Harvest by hand
2	Transfer grapes to winery	– Quick transportation – Low temperature
3	Destemming and crushing	– Removal of grapes that do not meet the quality criteria – We transfer the must with the grapes to the fermentation tanks
4	Pre-fermenting cryoextraction and first mud removal	– 6 h at 10–12 °C
5	Pressing grapes	– Separate outflow must from must after pressing in separate tanks
6	Second mud removal	– For a short period of 2–3 h
7	Alcoholic fermentation	– Temperature of fermentation tanks 20–22 °C – Aeration of the must by stirring the second and third day
8	Malolactic fermentation	– Malolactic fermentation in natural wine is almost sure
9	Third mud removal	– For a short period of 2–3 h
10	Fining wine	– Tanks must be completely full – After 2 months of wine stay in the tanks the wine must be bottled for greater protection against oxidation

Table 3.13 Soft Retsina favor, white wine protocol

	Steps	Additions/key points
1	Harvest	– Sampling and analyses – Low temperatures – Choose healthy grapes
2	Transfer grapes to winery	– Quick transportation – Low temperature
3	Destemming and crushing	– Removal of grapes that do not meet the quality criteria
4	Pre-fermenting cryoextraction	– 6 h at 15 °C – Addition of sulfur dioxide 100 g/tn – Addition of pectolytic enzymes specialized for this stage
5	Pressing grapes	– Separate outflow must from must after pressing in separate tanks
6	Mud removal	– Chemical analyses and corrections of must – Addition of pectolytic enzymes specialized for this stage (pectin esterase, polygalacturonase, and pectate lyase)
7	Alcoholic fermentation	– Addition of yeast nutrients – Addition of yeasts specialized for the grape variety – Addition of tannins selected for their antioxidant properties – Addition of glutathione formulation – Addition of pine resin 500 g per 1000 L of must – Temperature suitable based on yeast specifications – Aeration of the must by stirring the second and third day
8	Fining wine	– Chemical analyses and corrections – Removing clear wine from muds – Stabilizing produced wine (proteins, tartaric, calcium, color) – Removing bitters – Addition of Sulfur dioxide

3.10.5 Methods and Protocols for Making Retsina Wines

Retsina is a wine that, under the name Retsina, is legally only allowed to be produced in the Greek geographical area. While there is no legal prohibition regarding the color of the wine, it is customary for the resin to be a white wine to which pine resin is added during the alcoholic fermentation. The amount of resin varies according to the protocol, but in no case should it exceed one kilogram per hundred liters of must. It is a wine that differs, a special wine that characterizes the Greek vineyard. It is customary for Retsina wine to come from the "Savvatiano" and/or "Roditis" varieties. Because the aroma of pine resin is quite strong, it overshadows all the aromatic varietal characteristics, so it is not customary to use very aromatic grape varieties (Table 3.13).

3.10.6 Methods and Protocols for Making Wines with Different Fermentation Yeasts

The addition of yeasts to carry out alcoholic fermentation is done for many reasons. The most important reason is that we want to avoid spontaneous fermentation with an unknown native yeast strain. Another reason is that we want to speed up the start of alcoholic fermentation to reduce the risks of harmful bacteria growing. The addition of commercial yeasts is able to improve the aroma of the wine produced. In addition, with the selection of the yeast strain we can ensure repeatability in the final wine, which is very useful for the relationship of our wine with the consumer public.

In recent years, more and more wineries around the world are testing vinifications of the same must in separate tanks with different yeast strains. This is done either because they want to experiment with the final result or because they want to increase the versatility of their wines. We will present three different winemaking protocols for white, rosé, and red wines using different yeast strains. In any case it is very important to faithfully follow the rules of the protocol as well as the instructions given by the companies regarding the fermentation temperature which plays an important role in the final result (Tables 3.14, 3.15, and 3.16).

3.10.7 Methods and Protocols for Making Sweet Wines

This particular category is a huge separate category in the world of wines with a wide range and many lovers of the genre. The criterion for categorizing the wines of this category is their final sugar content. OIV has complementary definitions relating to sugar content as above. The wine is often described as:

- Dry: A wine is considered dry if it contains no more than 4 g/L of sugar or up to 9 g/L if the total acidity—measured in grams of tartaric acid per liter—is not more than 2 g/L lower than the sugar content.
- Medium dry: Medium dry wines are those where the sugar content exceeds the amount specified in the first bullet point but does not surpass 12 g/L, or 18 g/L if the difference between the sugar content and the total acidity level—expressed in grams per liter of tartaric acid—is no more than 10 g/L. This applies when the sugar content is greater than that in the first bullet point, up to a maximum of either 12 g/L or 18 g/L, provided the total acidity content is set as per the first bullet point.
- Semi-sweet: Semi-sweet wines are those with a sugar content exceeding the amount mentioned in the second bullet point, but not surpassing 45 g/L.
- Sweet, when the wine has a minimum sugar content of 45 g/L [17].

The production of sweet wines can be done in many ways that differ from region to region. In this section all known sweet winemaking protocols will be developed detailing for each stage of winemaking. The category of sweet wines also includes white and rosé and red wines as, as we said before, the specific category of wines depends on the final sugar content of the wines.

The most common way of producing sweet wines is to stop the alcoholic fermentation by cooling, sulfuring, and filtering with very small diameter filters which remove the yeasts that may in the future create re-fermentations of the residual sugars. Another way of producing sweet wines is to add concentrated grape must to dry wines.

Table 3.14 White wines protocol with different fermentation yeasts

	Steps	Additions/key points
1	Harvest	– Sampling and analyses – Low temperatures – Choose healthy grapes
2	Transfer grapes to winery	– Quick transportation – Low temperature
3	Destemming and crushing	– Removal of grapes that do not meet the quality criteria
4	Pre-fermenting cryoextraction	– 6 h at 15 °C – Addition of Sulfur dioxide 100 g/tn – Addition of pectolytic enzymes specialized for this stage
5	Pressing grapes	– Separate outflow must from must after pressing in two separate tanks and we name them outflow tank 1 and pressing tank 1
6	Mud removal	– Chemical analyses and corrections of must – Addition of pectolytic enzymes specialized for this stage (pectin esterase, polygalacturonase, and pectate lyase)
7	Separate fermentation tanks	– After the mud removal stage we transfer clear outflow must and clear pressing must from outflow tank 1 and pressing tank 1 to four different tanks as outflow tank 1A and outflow tank 1B and pressing tank 1A and pressing tank 1B
8	Alcoholic fermentation outflow tank 1A and pressing tank 1A	– Addition of yeast nutrients – Addition of first yeasts specialized for the grape variety in outflow tank 1A and pressing tank 1A – Addition of tannins selected for their antioxidant properties – Addition of glutathione formulation – Temperature suitable based on yeast specifications – Aeration of the must by stirring the second and third day
8	Alcoholic fermentation outflow tank 1B and pressing tank 1B	– Addition of yeast nutrients – Addition of second yeasts specialized for the grape variety in outflow tank 1B and pressing tank 1B – Addition of tannins selected for their antioxidant properties – Addition of glutathione formulation – Temperature suitable based on yeast specifications – Aeration of the must by stirring the second and third day
9	Fining wine	– Moving the wine with the same tank names – Chemical analyses and corrections – Removing clear wine from muds – Stabilizing produced wine (proteins, tartaric, calcium, color) – Removing bitters – Addition of Sulfur dioxide
10	Blending	– We are making organoleptic evaluation of each tank to check differences – We are making different blends to check the final result

Another special category of sweet wines known is known as mistela. Specifically during the grape harvest, we select specific vineyards to supply grapes for mistela. These grapes are left to ripen on the vine longer than those intended for wine production, resulting in higher sugar content for the mistela. On the day of harvest, the

Table 3.15 Rose wines protocol with different fermentation yeasts

	Steps	Additions/key points
1	Harvest	– Sampling and analyses – Low temperatures – Choose healthy grapes
2	Transfer grapes to winery	– Quick transportation – Low temperature
3	Destemming and crushing	– Removal of grapes that do not meet the quality criteria
4	Pre-fermenting cryoextraction	– 9 h at 13 °C – Addition of sulfur dioxide 120 g/tn – Addition of pectolytic enzymes specialized for this stage
5	Pressing grapes	– Separate outflow must from must after pressing in two separate tanks and we name them outflow tank 1 and pressing tank 1
6	Mud removal	– Chemical analyses and corrections of must – Addition of pectolytic enzymes specialized for this stage (pectin esterase, polygalacturonase, and pectate lyase)
7	Separate fermentation tanks	– After the mud removal stage we transfer clear outflow must and clear pressing must from outflow tank 1 and pressing tank 1 to four different tanks as outflow tank 1A and outflow tank 1B and pressing tank 1A and pressing tank 1B
8	Alcoholic fermentation outflow tank 1A and pressing tank 1A	– Addition of yeast nutrients – Addition of first yeasts specialized for the grape variety in outflow tank 1A and pressing tank 1A – Addition of tannins selected for their antioxidant properties – Addition of glutathione formulation – Temperature suitable based on yeast specifications – Aeration of the must by stirring the second and third day
8	Alcoholic fermentation outflow tank 1B and pressing tank 1B	– Addition of yeast nutrients – Addition of second yeasts specialized for the grape variety in outflow tank 1B and pressing tank 1B – Addition of tannins selected for their antioxidant properties – Addition of glutathione formulation – Temperature suitable based on yeast specifications – Aeration of the must by stirring the second and third day
9	Malolactic fermentation	– We recommend spontaneous malolactic fermentation without the addition of lactic acid bacteria. But if we notice that it is delayed then we add them
10	Fining wine	– Moving the wine with the same tank names – Chemical analyses and corrections – Removing clear wine from muds – Stabilizing produced wine (proteins, tartaric, calcium, color) – Removing bitters – Addition of Sulfur dioxide
10	Blending	– We are making organoleptic evaluation of each tank to check differences – We are making different blends to check the final result

Table 3.16 Red wines protocol with different fermentation yeasts

	Steps	Additions/key points
1	Harvest	– Sampling and analyses – Low temperatures – Choose healthy grapes
2	Transfer grapes to winery	– Quick transportation – Low temperature
3	Destemming and crushing	– Removal of grapes that do not meet the quality criteria – We transfer the must with the grapes to the fermentation tanks – Addition of Sulfur dioxide 100 g/tn – Addition of pectolytic enzymes specialized for this stage (maltodextrine, pectinases, ammonium sulfate)
4	Pre-fermenting cryoextraction	– Only in cases where we have excellent raw material
5	First mud removal	– Chemical analyses and corrections of must
5	Separate fermentation tanks	– After the mud removal stage we transfer must and grapes in two different fermentation tanks and we named them fermentation tank 1 and fermentation tank 2
8	Alcoholic fermentation tank 1	– Addition of yeast nutrients – Addition of first yeasts specialized for the grape variety in fermentation tank 1 – Addition of tannins selected for their antioxidant properties – Temperature suitable based on yeast specifications – Aeration of the must by stirring the second and third day
8	Alcoholic fermentation tank 2	– Addition of yeast nutrients – Addition of second yeasts specialized for the grape variety in fermentation tank 2 – Addition of tannins selected for their antioxidant properties – Temperature suitable based on yeast specifications – Aeration of the must by stirring the second and third day
5	Pressing grapes	– Moving wine to clear tanks for the completion of alcoholic fermentation with the same names tank 1 and tank 2
6	Completion of alcoholic fermentation	– Addition of yeast nutrients to each fermentation tank
7	Malolactic fermentation	– We recommend spontaneous malolactic fermentation without the addition of lactic acid bacteria. But if we notice that it is delayed then we add them
8	Second mud removal	– Chemical analyses and corrections of must
9	Fining wine	– Chemical analyses and corrections – Addition of Sulfur dioxide – Addition of polysaccharides – Organoleptic corrections – Stabilizing produced wine (proteins, tartaric, calcium, color) – Removing bitters
10	Blending	– We are making organoleptic evaluation of each tank to check differences – We are making different blends to check the final result

grapes are pressed, and an amount of wine alcohol is added to the extracted must to halt fermentation. This technique, called "mutage," preserves the grapes' natural sugars.

Another way of producing sweet wines is by naturally over-ripening the grapes. In particular, the cut grapes are spread or hung on specially designed clean surfaces to undergo natural drying through solar radiation. When the grapes acquire the desired sugar content, then the vinification process begins which due to the high sugar content will create a wine with a high alcoholic strength and the alcoholic fermentation will stop by itself. This is because the yeasts cannot function past a certain alcoholic strength.

For each of the above methods, three different vinification protocols are presented, one for white, one for rosé, and one for red wines. In all protocols, we should plan from the beginning their final sugar content. In our protocols, we aim to produce sweet wines with a final sugar content of more than 45 g/L (Tables 3.17, 3.18, 3.19, 3.20 and 3.21).

3.10.8 Methods and Protocols for Making Late Harvest Wines

Late harvest wines are produced from grapes that remain on the vine beyond their optimal ripeness. As the grapes hang, they gradually become sweeter, since dehydration concentrates their sugar content. These late harvest grapes, usually picked 40–60 days after the standard harvest, are utilized to create wines with both a higher residual sugar and a higher potential alcohol content compared to regular table wines.

To create this type of wine there are specific requirements regarding the location and the climatic conditions of the vineyard. This is true because the growth of the botrytis fungus needs specific climatic conditions so that its attack does not rot the grape but makes it like a raisin. This method is called noble rot and is not achieved simultaneously in all grapes. So we have consecutive harvests and separate small vinifications in order to ensure a uniform and correct result. It is a given that the harvest is carried out by hand by workers and the sugar content of the grapes affected by the fungus must definitely be greater than 260 g/L.

It is a method with a very high risk and of course a very small production per hectare that requires correct cultivation techniques to be able to ensure a correct result of the raw material. Because each region has its own particularity, we cannot give a basic instruction on how the grape will be ready for harvest, you must be in direct contact with the agronomist of your vineyard. The following protocol starts with the fact that we have grapes that are affected by the fungus Botrytis cinerea and their sugars content in must is more than 260 g/L (Table 3.22).

Table 3.17 Sweet white wine protocol by stopping alcoholic fermentation at the desired sugar levels

	Steps	Additions/key points
1	Harvest at 13.5 Baume or 24.36 ° Brix	– Sampling and analyses – Low temperatures – Choose healthy grapes
2	Transfer grapes to winery	– Quick transportation – Low temperature
3	Destemming and crushing	– Removal of grapes that do not meet the quality criteria
4	Pre-fermenting cryoextraction	– 6 h at 15 °C – Addition of Sulfur dioxide 100 g/tn – Addition of pectolytic enzymes specialized for this stage
5	Pressing grapes	– Moving must to clear tanks
6	Mud removal	– 3 h at 10 °C – Addition of pectolytic enzymes specialized for this stage (pectin esterase, polygalacturonase, and pectate lyase)
7	Alcoholic fermentation	– Addition of yeast nutrients – Addition of yeasts specialized for the grape variety – Addition of tannins selected for their antioxidant properties – Addition of glutathione formulation – Temperature suitable based on yeast specifications – Aeration of the must by stirring the second and third day – Chemical analyses and corrections of must special attention to total acidity and pH
8	Monitoring and stopping alcoholic fermentation	– If we want accurate measurements, we must perform analyses of reducing sugars. Otherwise we often count the Baume or the °Brix to stop the alcoholic fermentation. For a sweet wine above 45 g/L of sugars content we stop the alcoholic fermentation when we measure 2.8 Baume at 20 °C or 5 °Brix at the same temperature. If we want less sugar, we adjust our measurements. But we must know that with this method we have to intervene a little earlier than the desired level because the alcoholic fermentation does not stop immediately – In order to achieve our goal, quick interventions must be made – We lower the temperature of our cooling system to 5 °C – Add sulfuric anhydride 60 g/tn (if we add metabisulfite potassium we add 120 g/tn) – Add potassium sorbate 200 g/tn
9	– Filtering and removing clear wine from muds	– Moving wine to clear tanks by sterile filtration filter. For convenience we could first filter the wine with larger diameter filters and immediately repeat the filtering by sterile filtration filter
10	Fining wine	– Chemical analyses and corrections special attention to total acidity and pH – Stabilizing produced wine (proteins, tartaric, calcium, color) – Addition of Sulfur dioxide – This protocol gives us a wine of 11.0–11.5% alcohol and sugars over 45 g/L

Table 3.18 Sweet red wine protocol by stopping alcoholic fermentation at the desired sugar levels

	Steps	Additions/key points
1	Harvest 14.0 Baume or 25.27 °Brix	– Sampling and analyses – Low temperatures – Choose healthy grapes
2	Transfer grapes to winery	– Quick transportation – Low temperature
3	Destemming and crushing	– Removal of grapes that do not meet the quality criteria – We transfer the must with the grapes to the fermentation tanks – Addition of Sulfur dioxide 100 g/tn – Addition of pectolytic enzymes specialized for this stage (maltodextrine, pectinases, ammonium sulfate)
3*	Pre-fermenting cryoextraction	– Only in cases where we have excellent raw material
4	Alcoholic fermentation	– Addition of yeast nutrients – Addition of yeasts specialized for the grape variety – Addition of concentrated tannins proanthocyanidin type – Temperature suitable based on yeast specifications – Aeration of the must by stirring the second and third day – Monitoring alcoholic fermentation until the fermented must has 5.0 Baume or 8.95 °Brix
5	Pressing grapes	
6	Monitoring and stopping alcoholic fermentation	– Keep monitoring alcoholic fermentation procedure until we measure 2.8 Baume at 20 °C or 5 °Brix at the same temperature. If we want accurate measurements, we must perform analyses of reducing sugars. Otherwise we often count the Baume or the °Brix to stop the alcoholic fermentation. This protocol gives us a wine of 11.5–12.0% alcohol and sugars over 45 g/L. if we want less sugar content, we adjust our measurements. But we must know that with this method we have to intervene a little earlier than the desired level because the alcoholic fermentation does not stop immediately – In order to achieve our goal, quick interventions must be made – We lower the temperature of our cooling system to 5 °C – Add sulfuric anhydride 60 g/tn (if we add metabisulfite potassium we add 120 g/tn) – Add potassium sorbate 200 g/tn
7	– Filtering and removing clear wine from muds	– Moving wine to clear tanks by sterile filtration filter. For convenience we could first filter the wine with larger diameter filters and immediately repeat the filtering by sterile filtration filter
8	Fining wine	– Chemical analyses and corrections special attention to total acidity and pH – Addition of sulfur dioxide – Organoleptic corrections – Stabilizing produced wine (proteins, tartaric, calcium, color)

Table 3.19 Sweet rose wine protocol by stopping alcoholic fermentation at the desired sugar levels

	Steps	Additions/key points
1	Harvest at 13.5 Baume or 24.36 °Brix	– Sampling and analyses – Low temperatures – Choose healthy grapes
2	Transfer grapes to winery	– Quick transportation – Low temperature
3	Destemming and crushing	– Removal of grapes that do not meet the quality criteria
4	Pre-fermenting cryoextraction	– 6 h at 15 °C – Addition of Sulfur dioxide 100 g/tn – Addition of pectolytic enzymes specialized for this stage
5	Pressing grapes	– Moving grape must to clear tanks
6	Mud removal	– 3 h at 10 °C – Addition of pectolytic enzymes specialized for this stage (pectin esterase, polygalacturonase, and pectate lyase)
7	Alcoholic fermentation	– Chemical analyses and corrections special attention to total acidity and pH – Addition of yeast nutrients – Addition of yeasts specialized for the grape variety – Addition of tannins selected for their antioxidant properties – Temperature suitable based on yeast specifications – Aeration of the must by stirring the second and third day
8	Monitoring and stopping alcoholic fermentation	– If we want accurate measurements, we must perform analyses of reducing sugars. Otherwise we often count the Baume or the °Brix to stop the alcoholic fermentation. For a sweet wine above 45 g/L of sugars content we stop the alcoholic fermentation when we measure 2.8 Baume at 20 °C or 5 °Brix at the same temperature. If we want less sugar, we adjust our measurements. But we must know that with this method we have to intervene a little earlier than the desired level because the alcoholic fermentation does not stop immediately – In order to achieve our goal, quick interventions must be made – We lower the temperature of our cooling system to 5 °C – Add sulfuric anhydride 60 g/tn (if we add metabisulfite potassium we add 120 g/tn) – Add potassium sorbate 200 g/tn
9	– Filtering and removing clear wine from muds	– Moving wine to clear tanks by sterile filtration filter. For convenience we could first filter the wine with larger diameter filters and immediately repeat the filtering by sterile filtration filter
10	Fining wine	– Chemical analyses and corrections special attention to total acidity and pH – Stabilizing produced wine (proteins, tartaric, calcium, color) – Addition of Sulfur dioxide – This protocol gives us a wine of 11.0–11.5% alcohol and sugars over 45 g/L

Table 3.20 Mistela (Vin de liquer) protocol for white and rose wines

	Steps	Additions/key points
1	Harvest overripe grapes at 14.0 Baume or 25.27 °Brix	– Sampling and analyses – Low temperatures – Choose healthy grapes
2	Transfer grapes to winery	– Quick transportation – Low temperature
3	Destemming and crushing	– Removal of grapes that do not meet the quality criteria
4	Pre-fermenting cryoextraction	– 6 h at 5 °C – Addition of sulfur dioxide 100 g/tn – Addition of pectolytic enzymes specialized for this stage
5	Pressing grapes	– Moving must to clear tanks
6	Mud removal	– 3 h at 10 °C – Addition of pectolytic enzymes specialized for this stage (pectin esterase, polygalacturonase, and pectate lyase)
7	Filtering and removing clear wine from muds	– Moving wine to clear tanks by sterile filtration filter. For convenience we could first filter the wine with larger diameter filters and immediately repeat the filtering by sterile filtration filter
8	Alcohol addition	– Addition 16% of 96° % vol proof pure alcohol suitable for food
9	Fining wine	– Chemical analyses and corrections special attention to total acidity and pH – Stabilizing produced wine (proteins, tartaric, calcium, color) – Addition of Sulfur dioxide 60 g/tn – Add potassium sorbate 200 g/tn – This protocol gives us a wine of 16.0% alcohol and sugars over 200 g/L – Aging from 12 to 24 months in oak barrels is recommended

Table 3.21 Mistela (Vin de liquer) protocol for red wines

	Steps	Additions/key points
1	Harvest overripe grapes at 14.0 Baume or 25.27 °Brix	– Sampling and analyses – Low temperatures – Choose healthy grapes
2	Transfer grapes to winery	– Quick transportation – Low temperature
3	Destemming and crushing	– Removal of grapes that do not meet the quality criteria
4	Pre-fermenting cryoextraction	– 12 h at 5 °C – Frequent stirring of the crashed grapes and must – Addition of Sulfur dioxide 80 g/tn – Addition of pectolytic enzymes specialized for this stage
5	Pressing grapes	– Moving must to clear tanks
6	Mud removal	– 3 h at 10 °C – Addition of pectolytic enzymes specialized for this stage (pectin esterase, polygalacturonase, and pectate lyase)

(continued)

Table 3.21 (continued)

	Steps	Additions/key points
7	Filtering and removing clear wine from muds	– Moving wine to clear tanks by sterile filtration filter. For convenience we could first filter the wine with larger diameter filters and immediately repeat the filtering by sterile filtration filter
8	Alcohol addition	– Addition 16% of 96° % vol proof pure alcohol suitable for food
9	Fining wine	– Chemical analyses and corrections special attention to total acidity and pH – Stabilizing produced wine (proteins, tartaric, calcium, color) – Addition of sulfur dioxide 60 g/tn – Add potassium sorbate 200 g/tn – This protocol gives us a wine of 16.0% alcohol and sugars over 200 g/L – Aging from 12 to 36 months in oak barrels is recommended

Table 3.22 White, rose, and red late harvest wines protocol

	Steps	Additions/key points
1	Harvest at 14.5–15.0 Baume or 26.19–27.1 °Brix	– Sampling and analyses – Low temperatures – Choose grapes infected with fungus Botrytis cinerea
2	Transfer grapes to winery	– Quick transportation – Low temperature
3	Destemming and crushing	– Removal of grapes that do not meet the quality criteria
5	Pressing grapes	– Light pressure at the beginning that increases with the aim of the greater amount of must
6	Mud removal	– 6 h at 10 °C – Addition of pectolytic enzymes specialized for this stage (pectin esterase, polygalacturonase, and pectate lyase)
7	Alcoholic fermentation	– Addition of yeast nutrients – Addition of yeasts specialized for the grape variety – Addition of tannins selected for their antioxidant properties – Addition of glutathione formulation for white wines – Temperature suitable based on yeast specifications – Aeration of the must by stirring the second and third day – Chemical analyses and corrections of must special attention to total acidity and pH
8	Monitoring and stopping alcoholic fermentation	– If we want accurate measurements, we must perform analyses of reducing sugars. Otherwise we often count the Baume or the °Brix to stop the alcoholic fermentation. For a wine above 20 g/L of sugars content we stop the alcoholic fermentation when we measure 1.3 Baume at 20 °C or 2.32 °Brix at the same temperature. If we want less sugar, we adjust our measurements. But we must know that with this method we have to intervene a little earlier than the desired level because the alcoholic fermentation does not stop immediately – In order to achieve our goal, quick interventions must be made – We lower the temperature of our cooling system to 5 °C – Add sulfuric anhydride 75 g/tn (if we add metabisulfite potassium we add 150 g/tn) – Add potassium sorbate 200 g/tn

(continued)

Table 3.22 (continued)

	Steps	Additions/key points
9	Filtering and removing clear wine from muds	– Moving wine to clear tanks by sterile filtration filter. For convenience we could first filter the wine with larger diameter filters and immediately repeat the filtering by sterile filtration filter
10	Fining wine	– Chemical analyses and corrections special attention to total acidity and pH – Stabilizing produced wine (proteins, tartaric, calcium, color) – Addition of Sulfur dioxide – This protocol gives us a wine of 13.0–13.5% alcohol and sugars over 20 g/L

3.10.9 Methods and Protocols for Making Beaujolais Nouveau and Wines in CO_2 Environment

Beaujolais nouveau is a red wine produced from Gamay grapes produced in the Beaujolais region of France. It is a vin de primeur, fermented for just a few weeks before going on sale on the third Thursday of November [18]. Behind this beautiful wine story that has now become a tradition is a story of great technological interest. In this story biologists were studying the processes that take place in fruits when they are in an environment where there is no oxygen. These studies led to the observation of a metabolic activity without the intervention of exogenous microorganisms. Anaerobic fermentation metabolism is associated with the presence of the enzyme alcohol dehydrogenase in grapes [3]. This enzyme converts sugar into alcohol, functioning similarly to the alcohol dehydrogenase found in yeast. Intracellular fermentation results in the production of small quantities of ethanol. Additionally, by-products like glycerol and acetaldehyde accumulate in the berry. Organic nitrogen content undergoes changes, and there is a diffusion of phenolic and certain aromatic compounds from the skin into the pulp. Color compounds are extracted more easily than tannins. Catabolization of malic acid is another crucial process, leading to the formation of additional alcohol, succinic, and other acids, but not lactic acid [19]. Approximately 50% of the malic acid content can be converted in this manner, resulting in a pH increase in the berry of about 0.25. When alcohol production reaches approximately 2%, enzymatic activity stops, typically causing the berry to split and release its juice. The temperature of the fruit at harvest is critical, as it influences the rate of carbon dioxide absorption by the berries. Temperatures between 30 and 32 °C are likely to yield the best-structured wines, although the duration of the carbonic maceration phase can vary depending on the desired wine style. When we have low temperatures during harvest the grapes absorb less carbon dioxide and take a long time to complete the fermentation process in the fruit, also producing a wine much lower in aroma (Table 3.23).

Table 3.23 Beaujolais nouveau for white, rose and red wines protocol

	Steps	Additions/key points
1	Harvest 14.0 Baume or 25.27 °Brix at 30–33 °C	– Sampling and analyses – Low temperatures – Choose healthy grapes
2	Transfer grapes to winery	– Quick transportation – Low temperature
3	Preparation of fermentation tank	– A small amount of grape must in the fermentation tank which is in alcoholic fermentation to produce carbon dioxide – The fermentation tank must have some pressure resistance – Addition of yeast nutrients – Addition of yeasts specialized for the grape variety – Addition of concentrated tannins proanthocyanidin type
4	Filling fermentation tank with unstemmed grapes	– Unstemmed grapes are manually transferred to the fermentation tank containing the under-fermented must. Fill the tank with the grapes and close it. We frequently check the pressure relief valve to see that it is working properly to avoid accidents from accumulated pressure – This stage lasts at temperatures of around 30° from 7 to 10 days, depending on the characteristics we want to give to our wine
5	Pressing grapes	
6	Alcoholic fermentation	– Addition of yeast nutrients – Addition of concentrated tannins proanthocyanidin type – Temperature suitable based on yeast specifications – Aeration of the must by stirring the second and third day – Completing alcoholic fermentation
7	Malolactic fermentation	– We recommend spontaneous malolactic fermentation without the addition of lactic acid bacteria. But if we notice that it is delayed then we add them
8	– Filtering and removing clear wine from muds	– Moving wine to clear tanks by sterile filtration filter. For convenience we could first filter the wine with larger diameter filters and immediately repeat the filtering by sterile filtration filter
9	Fining wine	– Chemical analyses and corrections special attention to total acidity and pH – Addition of Sulfur dioxide – Organoleptic corrections – Stabilizing produced wine (proteins, tartaric, calcium, color) – Bottling wines at the end of October

3.10.10 Methods and Protocols for Making Sherry Wines

Sherry wines are a type of wine that characterizes the Spanish vineyard and is a quality wine that is governed by strict legislation. It is a wine with a long past and certainly a long future. Sherry is a fortified wine made from white grapes that are grown near the city of Jerez de la Frontera in Andalusia. Sherry is a fortified wine available in various styles, predominantly made from the Palomino grape. It ranges from lighter types akin to white table wines, like Manzanilla and Fino, to darker, more robust versions that undergo oxidation during barrel aging, such as Amontillado and Oloroso [20].

The most characteristic stage in the production of sherry wines takes place during the aging in oak barrels for a period of 12–36 months. The oak barrels are filled by about 1/6. A special fungus called flor is created on the surface of the wine which can withstand the percentage of alcohol that our wine has. In addition to specific alcohol levels, flor requires a very particular climate to thrive. Humidity is a crucial factor, ideally at 70% or higher. This fungus creates a thick layer on the surface of the wine so that it does not come into contact with oxygen, thus preventing its oxidation and allowing its biological maturation. The sherry wines produced with biological maturation named sherry Fino [21].

Sometimes biological and oxidative maturation are observed simultaneously during the aging process in the wine, the resulting wine is darker than fino wine and is called Amondillado [22]. Sherry wines produced by oxidation maturation called Oloroso this wines before entering oak barrels is strengthened up to 17.5° of alcohol to prevent the growth of the flor fungus (Table 3.24) [22].

Table 3.24 Sherry white wines protocol

	Steps	Additions/key points
1	Harvest overripe grapes at 13.0 Baume or 23.45 °Brix	– Sampling and analyses – Low temperatures – Choose healthy grapes
2	Transfer grapes to winery	– Quick transportation – Low temperature
3	Destemming and crushing	– Removal of grapes that do not meet the quality criteria
4	Pre-fermenting cryoextraction	– 12 h at 5 °C – Frequent stirring of the crashed grapes and must – Addition of Sulfur dioxide 60 g/tn – Addition of pectolytic enzymes specialized for this stage
5	Pressing grapes	– Moving must to clear tanks
6	Mud removal	– 3 h at 10 °C – Addition of pectolytic enzymes specialized for this stage (pectin esterase, polygalacturonase, and pectate lyase)
7	Alcoholic fermentation	– Addition of yeast nutrients – Addition of concentrated tannins proanthocyanidin type – Temperature suitable based on yeast specifications – Aeration of the must by stirring the second and third day – Completing alcoholic fermentation
7	Filtering and removing clear wine from muds	– Moving wine to clear tanks by filtration
8	Alcohol addition	– Addition 96° % vol proof pure alcohol suitable for food until 15.5% vol wine for Fino sherry or 17.5% vol to produce Oloroso sherry
9	Oak barrels	– Chemical analyses and corrections special attention to total acidity and pH – Addition of Sulfur dioxide 60 g/tn – Aging 36 months in oak barrels is recommended – The barrels must be at least 1/6 empty – Read above for instruction to produce sherry Fino, Amondillado, or Oloroso

References

1. de Castilhos MBM (2023) Basic protocols in enology and winemaking, methods and protocols in food science. Springer Protocols. Humana, New York. https://doi.org/10.1007/978-1-0716-3088-4
2. Storm DR (2013) Winery utilities, planning, design and operation. Springer, New York. https://doi.org/10.1007/978-1-4757-6973-9
3. Jastrzębska A (2022) Determination of home-made wine selected parameters and study of honey addition impact on pro-healthy components content. Eur Food Res Technol 248:963–973. https://doi.org/10.1007/s00217-021-03934-4
4. Płotka-Wasylka J, Simeonov V, Namieśnik J (2018) Characterization of home-made and regional fruit wines by evaluation of correlation between selected chemical parameters. Microchem J 140:66–73. https://doi.org/10.1016/j.microc.2018.04.010
5. Sneha N, Sundaram M, Ranjan R (2024) Acre-scale grape bunch detection and predict grape harvest using YOLO deep learning network. SN Comput Sci 5:250. https://doi.org/10.1007/s42979-023-02572-9
6. Camps JO, Ramos MC (2012) Grape harvest and yield responses to inter-annual changes in temperature and precipitation in an area of North-East Spain with a Mediterranean climate. Int J Biometeorol 56:853–864. https://doi.org/10.1007/s00484-011-0489-3
7. Caprio JM, Quamme HA (2002) Weather conditions associated with grape production in the Okanagan Valley of British Columbia and potential impact of climate change. Can J Plant Sci 82:755–763
8. Coombe BG (1987) Influence of temperature on composition and quality of grapes. Acta Hortic 206:23–36
9. de Orduña RM (2010) Climate change associated effects on grape and wine quality and production. Food Res Int 43:1844–1855
10. Darnal A, Poggesi S, Ceci AT et al (2023) Interactive effect of pre-fermentative grape freezing and malolactic fermentation on the anthocyanins profile in red wines prone to colour instability. Eur Food Res Technol 249:2045–2065. https://doi.org/10.1007/s00217-023-04270-5
11. Sacchi KL, Bisson LF, Adams DO (2005) A review of the effect of winemaking techniques on phenolic extraction in red wines. Am J Enol Vitic 56(3):197–206
12. Kersh DME, Hammad G, Donia MS et al (2023) A comprehensive review on grape juice beverage in context to its processing and composition with future perspectives to maximize its value. Food Bioprocess Technol 16:1–23. https://doi.org/10.1007/s11947-022-02858-5
13. Varela F, Calderón F, González M et al (1999) Effect of clarification on the fatty acid composition of grape must and the fermentation kinetics of white wines. Eur Food Res Technol 209:439–444. https://doi.org/10.1007/s002170050523
14. Boulton RB, Singleton VL, Bisson LF, Kunkee RE (1999) The fining and clarification of wines. In: Principles and practices of winemaking. Springer, Boston. https://doi.org/10.1007/978-1-4757-6255-6_7
15. Lee FA (1983) Alcoholic fermentation. In: Basic food chemistry. Springer, Dordrecht. https://doi.org/10.1007/978-94-011-7376-6_14
16. Binod P, Sindhu R, Pandey A (2013) The alcohol fermentation step: the most common ethanologenic microorganisms among yeasts, bacteria and filamentous fungi. In: Faraco V (ed) Lignocellulose conversion. Springer, Berlin. https://doi.org/10.1007/978-3-642-37861-4_7
17. Blade HW, Boulton RB (1988) Adsorption of protein by bentonite in a model wine solution. Am J Enol Vitic 39:193–199
18. Somers TC, Vérette E (1988) Phenolic composition of natural wine types. In: Linskens HF, Jackson JF (eds) Wine analysis, modern methods of plant analysis, vol 6. Springer, Berlin. https://doi.org/10.1007/978-3-642-83340-3_8
19. Ruffner HP, Possner D, Brem S et al (1984) The physiological role of malic enzyme in grape ripening. Planta 160:444–448. https://doi.org/10.1007/BF00429761

20. Reader HP, Dominguez M (1995) Fortified wines: sherry, port and Madeira. In: Lea AGH, Piggott JR (eds) Fermented beverage production. Springer, Boston. https://doi.org/10.1007/978-1-4757-5214-4_7
21. van der Schee HA, Bouwknegt JP, Tas AC et al (1989) The authentication of sherry wines using pattern recognition: an inter laboratory study. Z Lebensm Unters Forch 188:324–329. https://doi.org/10.1007/BF01352390
22. Ribereau-gayon P, Dubourdieu D, Donèche B, Lonvaud A (2000) Handbook of enology, vol 1: the microbiology of wine and vinifications. Wiley, Hoboken

Wine Analysis Methods

4

4.1 Sugars

As the composition of the must has been extensively analyzed in Chap. 2, from a chemical point of view, the must is an aqueous solution of various organic and inorganic substances that are components of the grape. The most important of these for winemaking are sugars. The grape must is examined for its suitability for vinification as well as any corrections before the vinification process begins. Determination of sugars is important.

4.1.1 Evaluation by Refractometry of the Sugar Concentration in Grape Musts

Principle
The refractive index at 20 °C, whether expressed as an absolute value or as a percentage by mass of sucrose, is provided in the relevant table. This facilitates the determination of sugar concentration in grape musts, concentrated grape musts, and rectified concentrated grape musts in both grams per liter and grams per kilogram.

Apparatus

1. **Abbe refractometer**
   The refractometer used must be fitted with a scale giving:
   - Either percentage by mass of sucrose to 0.1%;
   - Or refractive indices to four decimal places.
2. **Thermometer**
   The refractometer must be equipped with a thermometer having a scale extending at least from +15 °C to +25 °C and with a system for circulating water that will enable measurements to be made at a temperature of 20 ± 5 °C.

G. Z. Kyzas, F. Papageorgiou, *Oenology in Practice*,
https://doi.org/10.1007/978-3-031-85531-3_4

3. **Absorbent paper**
4. **Dropper**
5. **Grape must**
6. **Distilled water**

Preparation of the Grape Must Sample
The grape must we are going to measure should be free from suspensions. If we deem it necessary, we can filter the measured must by a lab filter or dry gauze.

Procedure
Bring the sample to a temperature as close as we can to 20 °C. We check if the refractometer shows zero when we place distilled water at a temperature of 20 on the surface of the prism, if necessary we make the necessary correction. Then we wipe the surface of the prism very well and with the use of the dropper, we place a drop of the must under examination. We coat the drop well on the surface of the prism and close the moving part of it. Then we point the refractometer in the direction of a light source to read the result. The reading is achieved by moving the microscrew of the device so that a clear dividing line appears between the bright and dark surface of the field of view.

The result can be expressed as either the percentage by mass of sucrose to the nearest 0.1 or the refractive index to four decimal places. Perform at least two determinations on the same prepared sample. Note the temperature t °C.

Calculation

Temperature Correction

- For Abbe refractometer calibrated in percentage by mass of sucrose: refer to Table 4.1 for the temperature correction.
- For Abbe refractometer calibrated in refractive index: locate the index measured at t °C in Table 4.2 (Column 2) to determine the equivalent percentage by mass of Saccharose at t °C (Column 1). This figure is then adjusted for temperature and represented as a concentration at 20 °C using Table 4.1.

Sugar Concentration in Must and Concentrated Must
Find the percentage by mass of sucrose at 20 °C in Table 4.2 and read from the same row the sugar concentration in grams per liter and grams per kilogram. The sugar concentration is expressed in terms of invert sugar to one decimal place.

Sugar Concentration in Rectified Concentrated Must
Find the percentage by mass of sucrose at 20 °C in Table 4.3 and read from the same row the sugar concentration in grams per liter and grams per kilogram. The sugar concentration is expressed in terms of invert sugar to one decimal place.

Table 4.1 Correction to be made in the case where the percentage by mass of saccharose was determined at a temperature different by 20 °C

Temperature	Percentage by mass measured in %													
	10	15	20	25	30	35	40	45	50	55	60	65	70	75
5	−0.82	−0.87	−0.92	−0.95	−0.99									
6	−0.8	−0.82	−0.87	−0.9	−0.94									
7	−0.74	−0.78	−0.82	−0.84	−0.88									
8	−0.69	−0.73	−0.76	−0.79	−0.82									
9	−0.64	−0.67	−0.71	−0.73	−0.75									
10	−0.59	−0.62	−0.65	−0.67	−0.69	−0.71	−0.72	−0.73	−0.74	−0.75	−0.75	−0.75	−0.75	−0.75
11	−0.54	−0.57	−0.59	−0.61	−0.63	−0.64	−0.65	−0.66	−0.67	−0.68	−0.68	−0.68	−0.68	−0.67
12	−0.49	−0.51	−0.53	−0.55	−0.56	−0.57	−0.58	−0.59	−0.6	−0.6	−0.61	−0.61	−0.6	−0.6
13	−0.43	−0.45	−0.47	−0.48	−0.5	−0.51	−0.52	−0.52	−0.53	−0.53	−0.53	−0.53	−0.53	−0.53
14	−0.38	−0.39	−0.4	−0.42	−0.43	−0.44	−0.44	−0.45	−0.45	−0.46	−0.46	−0.46	−0.46	−0.45
15	−0.32	−0.33	−0.34	−0.35	−0.36	−0.37	−0.37	−0.38	−0.38	−0.38	−0.38	−0.38	−0.38	−0.38
16	−0.26	−0.27	−0.28	−0.28	−0.29	−0.3	−0.3	−0.3	−0.31	−0.31	−0.31	−0.31	−0.31	−0.3
17	−0.2	−0.2	−0.21	−0.21	−0.22	−0.22	−0.23	−0.23	−0.23	−0.23	−0.23	−0.23	−0.23	−0.23
18	−0.13	−0.14	−0.14	−0.14	−0.15	−0.15	−0.15	−0.15	−0.15	−0.15	−0.15	−0.15	−0.15	−0.15
19	−0.07	−0.07	−0.07	−0.07	−0.07	−0.08	−0.08	−0.08	−0.08	−0.08	−0.08	−0.08	−0.08	−0.08
20	Reference													
21	0.07	0.07	0.07	0.07	0.08	0.08	0.08	0.08	0.08	0.08	0.08	0.08	0.08	0.08
22	0.14	0.14	0.15	0.15	0.15	0.15	0.16	0.16	0.16	0.16	0.16	0.16	0.15	0.15
23	0.21	0.22	0.22	0.23	0.23	0.23	0.23	0.24	0.24	0.24	0.24	0.23	0.23	0.23
24	0.29	0.29	0.3	0.3	0.31	0.31	0.31	0.32	0.32	0.32	0.32	0.31	0.31	0.31
25	0.36	0.37	0.38	0.38	0.39	0.39	0.4	0.4	0.4	0.4	0.4	0.39	0.39	0.39
26	0.44	0.45	0.46	0.46	0.47	0.47	0.48	0.48	0.48	0.48	0.48	0.47	0.47	0.46
27	0.52	0.53	0.54	0.55	0.55	0.56	0.56	0.56	0.56	0.56	0.56	0.55	0.55	0.54
28	0.6	0.61	0.62	0.63	0.64	0.64	0.64	0.65	0.65	0.64	0.64	0.64	0.63	0.62

(continued)

Table 4.1 (continued)

Temperature	Percentage by mass measured in %													
29	0.68	0.69	0.7	0.71	0.72	0.73	0.73	0.73	0.73	0.73	0.72	0.72	0.71	0.7
30	0.77	0.78	0.79	0.8	0.81	0.81	0.81	0.82	0.81	0.81	0.81	0.8	0.79	0.78
31	0.85	0.87	0.88	0.89	0.89	0.9	0.9	0.9	0.9	0.9	0.89	0.88	0.87	0.86
32	0.94	0.95	0.96	0.97	0.98	0.99	0.99	0.99	0.99	0.98	0.97	0.96	0.95	0.94
33	1.03	1.04	1.05	1.06	1.07	1.08	1.08	1.08	1.07	1.07	1.06	1.05	1.03	1.02
34	1.12	1.19	1.15	1.15	1.16	1.17	1.17	1.17	1.16	1.15	1.14	1.13	1.12	1.1
35	1.22	1.23	1.24	1.25	1.25	1.26	1.26	1.25	1.25	1.24	1.23	1.21	1.2	1.18
36	1.31	1.32	1.33	1.34	1.35	1.35	1.35	1.35	1.34	1.33	1.32	1.3	1.28	1.26
37	1.41	1.42	1.43	1.44	1.44	1.44	1.44	1.44	1.43	1.42	1.4	1.38	1.36	1.34
38	1.51	1.52	1.53	1.53	1.54	1.54	1.53	1.53	1.52	1.51	1.49	1.47	1.45	1.42
39	1.61	1.62	1.62	1.63	1.63	1.63	1.63	1.62	1.61	1.6	1.58	1.56	1.53	1.5
40	1.71	1.72	1.72	1.73	1.73	1.73	1.72	1.71	1.7	1.69	1.67	1.64	1.62	1.59

Table 4.2 The sugar content of musts and concentrated musts in grams per liter and in grams per kilogram, determined using a graduated refractometer, either in percentage by mass of saccharose at 20 °C or refractive index at 20 °C. The mass density at 20 °C is also given

Saccharose % (m/m).	Refractive index at 20 °C	Mass density at 20 °C	Sugars in g/L	Sugars in g/kg	ABV % vol at 20 °C
10.0	1.34782	1.0391	82.2	79.1	4.89
10.1	1.34798	1.0395	83.3	80.1	4.95
10.2	1.34813	1.0399	84.3	81.1	5.01
10.3	1.34829	1.0403	85.4	82.1	5.08
10.4	1.34844	1.0407	86.5	83.1	5.14
10.5	1.34860	1.0411	87.5	84.1	5.20
10.6	1.34875	1.0415	88.6	85.0	5.27
10.7	1.34891	1.0419	89.6	86.0	5.32
10.8	1.34906	1.0423	90.7	87.0	5.39
10.9	1.34922	1.0427	91.8	88.0	5.46
11.0	1.34937	1.0431	92.8	89.0	5.52
11.1	1.34953	1.0436	93.9	90.0	5.58
11.2	1.34968	1.0440	95.0	91.0	5.65
11.3	1.34984	1.0444	96.0	92.0	5.71
11.4	1.34999	1.0448	97.1	92.9	5.77
11.5	1.35015	1.0452	98.2	93.9	5.84
11.6	1.35031	1.0456	99.3	94.9	5.90
11.7	1.35046	1.0460	100.3	95.9	5.96
11.8	1.35062	1.0464	101.4	96.9	6.03
11.9	1.35077	1.0468	102.5	97.9	6.09
12.0	1.35093	1.0472	103.5	98.9	6.15
12.1	1.35108	1.0477	104.6	99.9	6.22
12.2	1.35124	1.0481	105.7	100.8	6.28
12.3	1.35140	1.0485	106.8	101.8	6.35
12.4	1.35156	1.0489	107.8	102.8	6.41
12.5	1.35171	1.0493	108.9	103.8	6.47
12.6	1.35187	1.0497	110.0	104.8	6.54
12.7	1.35203	1.0501	111.1	105.8	6.60
12.8	1.35219	1.0506	112.2	106.8	6.67
12.9	1.35234	1.0501	113.2	107.8	6.73
13.0	1.35250	1.0514	114.3	108.7	6.79
13.1	1.35266	1.0518	115.4	109.7	6.86
13.2	1.35282	1.0522	116.5	110.7	6.92
13.3	1.35298	1.0527	117.6	111.7	6.99
13.4	1.35313	1.0531	118.7	112.7	7.05
13.5	1.35329	1.0535	119.7	113.7	7.11
13.6	1.35345	1.0539	120.8	114.7	7.18
13.7	1.35361	1.0543	121.9	115.6	7.24
13.8	1.35377	1.0548	123.0	116.6	7.31
13.9	1.35393	1.0552	124.1	117.6	7.38
14.0	1.35408	1.0556	125.2	118.6	7.44

(continued)

Table 4.2 (continued)

Saccharose % (m/m).	Refractive index at 20 °C	Mass density at 20 °C	Sugars in g/L	Sugars in g/kg	ABV % vol at 20 °C
14.1	1.35424	1.0560	126.3	119.6	7.51
14.2	1.35440	1.0564	127.4	120.6	7.57
14.3	1.35456	1.0569	128.5	121.6	7.64
14.4	1.35472	1.0573	129.6	122.5	7.70
14.5	1.35488	1.0577	130.6	123.5	7.76
14.6	1.35504	1.0581	131.7	124.5	7.83
14.7	1.35520	1.0586	132.8	125.5	7.89
14.8	1.35536	1.0590	133.9	126.5	7.96
14.9	1.35552	1.0594	135.0	127.5	8.02
15.0	1.35568	1.0598	136.1	128.4	8.09
15.1	1.35584	1.0603	137.2	129.4	8.15
15.2	1.35600	1.0607	138.3	130.4	8.22
15.3	1.35616	1.0611	139.4	131.4	8.28
15.4	1.35632	1.0616	140.5	132.4	8.35
15.5	1.35648	1.0620	141.6	133.4	8.42
15.6	1.35664	1.0624	142.7	134.3	8.48
15.7	1.35680	1.0628	143.8	135.3	8.55
15.8	1.35696	1.0633	144.9	136.3	8.61
15.9	1.35713	1.0637	146.0	137.3	8.68
16.0	1.35729	1.0641	147.1	138.3	8.74
16.1	1.35745	1.0646	148.2	139.3	8.81
16.2	1.35761	1.0650	149.3	140.2	8.87
16.3	1.35777	1.0654	150.5	141.2	8.94
16.4	1.35793	1.0659	151.6	142.2	9.01
16.5	1.35810	1.0663	152.7	143.2	9.07
16.6	1.35826	1.0667	153.8	144.2	9.14
16.7	1.35842	1.0672	154.9	145.1	9.21
16.8	1.35858	1.0676	156.0	146.1	9.27
16.9	1.35874	1.0680	157.1	147.1	9.34
17.0	1.35891	1.0685	158.2	148.1	9.40
17.1	1.35907	1.0689	159.3	149.1	9.47
17.2	1.35923	1.0693	160.4	150.0	9.53
17.3	1.35940	1.0698	161.6	151.0	9.60
17.4	1.35956	1.0702	162.7	152.0	9.67
17.5	1.35972	1.0707	163.8	153.0	9.73
17.6	1.35989	1.0711	164.9	154.0	9.80
17.7	1.36005	1.0715	166.0	154.9	9.87
17.8	1.36021	1.0720	167.1	155.9	9.93
17.9	1.36038	1.0724	168.3	156.9	10.00
18.0	1.36054	1.0729	169.4	157.9	10.07
18.1	1.36070	1.0733	170.5	158.9	10.13
18.2	1.36087	1.0737	171.6	159.8	10.20
18.3	1.36103	1.0742	172.7	160.8	10.26

(continued)

Table 4.2 (continued)

Saccharose % (m/m).	Refractive index at 20 °C	Mass density at 20 °C	Sugars in g/L	Sugars in g/kg	ABV % vol at 20 °C
18.4	1.36120	1.0746	173.9	161.8	10.33
18.5	1.36136	1.0751	175.0	162.8	10.40
18.6	1.36153	1.0755	176.1	163.7	10.47
18.7	1.36169	1.0760	177.2	164.7	10.53
18.8	1.36185	1.0764	178.4	165.7	10.60
18.9	1.36202	1.0768	179.5	166.7	10.67
19.0	1.36219	1.0773	180.6	167.6	10.73
19.1	1.36235	1.0777	181.7	168.6	10.80
19.2	1.36252	1.0782	182.9	169.6	10.87
19.3	1.36268	1.0786	184.0	170.6	10.94
19.4	1.36285	1.0791	185.1	171.5	11.00
19.5	1.36301	1.0795	186.2	172.5	11.07
19.6	1.36318	1.0800	187.4	173.5	11.14
19.7	1.36334	1.0804	188.5	174.5	11.20
19.8	1.36351	1.0809	189.6	175.4	11.27
19.9	1.36368	1.0813	190.8	176.4	11.34
20.0	1.36384	1.0818	191.9	177.4	11.40
20.1	1.36401	1.0822	193.0	178.4	11.47
20.2	1.36418	1.0827	194.2	179.3	11.54
20.3	1.36434	1.0831	195.3	180.3	11.61
20.4	1.36451	1.0836	196.4	181.3	11.67
20.5	1.36468	1.0840	197.6	182.3	11.74
20.6	1.36484	1.0845	198.7	183.2	11.81
20.7	1.36501	1.0849	199.8	184.2	11.87
20.8	1.36518	1.0854	201.0	185.2	11.95
20.9	1.36535	1.0858	202.1	186.1	12.01
21.0	1.36551	1.0863	203.3	187.1	12.08
21.1	1.36568	1.0867	204.4	188.1	12.15
21.2	1.36585	1.0872	205.5	189.1	12.21
21.3	1.36602	1.0876	206.7	190.0	12.28
21.4	1.36619	1.0881	207.8	191.0	12.35
21.5	1.36635	1.0885	209.0	192.0	12.42
21.6	1.36652	1.0890	210.1	192.9	12.49
21.7	1.36669	1.0895	211.3	193.9	12.56
21.8	1.36686	1.0899	212.4	194.9	12.62
21.9	1.36703	1.0904	213.6	195.9	12.69
22.0	1.36720	1.0908	214.7	196.8	12.76
22.1	1.36737	1.0913	215.9	197.8	12.83
22.2	1.36754	1.0917	217.0	198.8	12.90
22.3	1.36771	1.0922	218.2	199.7	12.97
22.4	1.36787	1.0927	219.3	200.7	13.03
22.5	1.36804	1.0931	220.5	201.7	13.10
22.6	1.36821	1.0936	221.6	202.6	13.17

(continued)

Table 4.2 (continued)

Saccharose % (m/m).	Refractive index at 20 °C	Mass density at 20 °C	Sugars in g/L	Sugars in g/kg	ABV % vol at 20 °C
22.7	1.36838	1.0940	222.8	203.6	13.24
22.8	1.36855	1.0945	223.9	204.6	13.31
22.9	1.36872	1.0950	225.1	205.5	13.38
23.0	1.36889	1.0954	226.2	206.5	13.44
23.1	1.36906	1.0959	227.4	207.5	13.51
23.2	1.36924	1.0964	228.5	208.4	13.58
23.3	1.36941	1.0968	229.7	209.4	13.65
23.4	1.36958	1.0973	230.8	210.4	13.81
23.5	1.36975	1.0977	232.0	211.3	13.79
23.6	1.36992	1.0982	233.2	212.3	13.86
23.7	1.37009	1.0987	234.3	213.3	13.92
23.8	1.37026	1.0991	235.5	214.2	14.00
23.9	1.37043	1.0996	236.6	215.2	14.06
24.0	1.37060	1.1001	237.8	216.2	14.13
24.1	1.37078	1.1005	239.0	217.1	14.20
24.2	1.37095	1.1010	240.1	218.1	14.27
24.3	1.37112	1.1015	241.3	219.1	14.34
24.4	1.37129	1.1019	242.5	220.0	14.41
24.5	1.37146	1.1024	243.6	221.0	14.48
24.6	1.37164	1.1029	244.8	222.0	14.55
24.7	1.37181	1.1033	246.0	222.9	14.62
24.8	1.37198	1.1038	247.1	223.9	14.69
24.9	1.37216	1.1043	248.3	224.8	14.76
25.0	1.37233	1.1047	249.5	225.8	14.83
25.1	1.37250	1.1052	250.6	226.8	14.89
25.2	1.37267	1.1057	251.8	227.7	14.96
25.3	1.37285	1.1062	253.0	228.7	15.04
25.4	1.37302	1.1066	254.1	229.7	15.10
25.5	1.37319	1.1071	255.3	230.6	15.17
25.6	1.37337	1.1076	256.5	231.6	15.24
25.7	1.37354	1.1080	257.7	232.5	15.32
25.8	1.37372	1.1085	258.8	233.5	15.38
25.9	1.37389	1.1090	260.0	234.5	15.45
26.0	1.37407	1.1095	261.2	235.4	15.52
26.1	1.37424	1.1099	262.4	236.4	15.59
26.2	1.37441	1.1104	263.6	237.3	15.67
26.3	1.37459	1.1109	264.7	238.3	15.73
26.4	1.37476	1.1114	265.9	239.3	15.80
26.5	1.37494	1.1118	267.1	240.2	15.87
26.6	1.37511	1.1123	268.3	241.2	15.95
26.7	1.37529	1.1128	269.5	242.1	16.02
26.8	1.37546	1.1133	270.6	243.1	16.08
26.9	1.37564	1.1138	271.8	244.1	16.15

(continued)

Table 4.2 (continued)

Saccharose % (m/m).	Refractive index at 20 °C	Mass density at 20 °C	Sugars in g/L	Sugars in g/kg	ABV % vol at 20 °C
27.0	1.37582	1.1142	273.0	245.0	16.22
27.1	1.37599	1.1147	274.2	246.0	16.30
27.2	1.37617	1.1152	275.4	246.9	16.37
27.3	1.37634	1.1157	276.6	247.9	16.44
27.4	1.37652	1.1161	277.8	248.9	16.51
27.5	1.37670	1.1166	287.9	249.8	16.58
27.6	1.37687	1.1171	280.1	250.8	16.65
27.7	1.37705	1.1176	281.3	251.7	16.72
27.8	1.37723	1.1181	282.5	252.7	16.79
27.9	1.37740	1.1185	283.7	253.6	16.86
28.0	1.37758	1.1190	284.9	254.6	16.93
28.1	1.37776	1.1195	286.1	255.5	17.00
28.2	1.37793	1.1200	287.3	256.5	17.07
28.3	1.37811	1.1205	288.5	257.5	17.15
28.4	1.37829	1.1210	289.7	258.4	17.22
28.5	1.37847	1.1214	290.9	259.4	17.29
28.6	1.37864	1.1219	292.1	260.3	17.36
28.7	1.37882	1.1224	293.3	261.3	17.43
28.8	1.37900	1.1229	294.5	262.2	17.50
28.9	1.37918	1.1234	295.7	263.2	17.57
29.0	1.37936	1.1239	296.9	264.2	17.64
29.1	1.37954	1.1244	298.1	265.1	17.72
29.2	1.37972	1.1248	299.3	266.1	17.79
29.3	1.37989	1.1253	300.5	267.0	17.86
29.4	1.38007	1.1258	301.7	268.0	17.93
29.5	1.38025	1.1263	302.9	268.9	18.00
29.6	1.38043	1.1268	304.1	269.9	18.07
29.7	1.38061	1.1273	305.3	270.8	18.14
29.8	1.38079	1.1278	306.5	271.8	18.22
29.9	1.38097	1.1283	307.7	272.7	18.29
30.0	1.38115	1.1287	308.9	273.7	18.36
30.1	1.38133	1.1292	310.1	274.6	18.43
30.2	1.38151	1.1297	311.3	275.6	18.50
30.3	1.38169	1.1302	312.6	276.5	18.58
30.4	1.38187	1.1307	313.8	277.5	18.65
30.5	1.38205	1.1312	315.0	278.5	18.72
30.6	1.38223	1.1317	316.2	279.4	18.79
30.7	1.38241	1.1322	317.4	280.4	18.86
30.8	1.38259	1.1327	318.6	281.3	18.93
30.9	1.38277	1.1332	319.8	282.3	19.01
31.0	1.38296	1.1337	321.1	283.2	19.08
31.1	1.38314	1.1342	322.3	284.2	19.15
31.2	1.38332	1.1346	323.5	285.1	19.23

(continued)

Table 4.2 (continued)

Saccharose % (m/m).	Refractive index at 20 °C	Mass density at 20 °C	Sugars in g/L	Sugars in g/kg	ABV % vol at 20 °C
31.3	1.38350	1.1351	324.7	286.1	19.30
31.4	1.38368	1.1356	325.9	287.0	19.37
31.5	1.38386	1.1361	327.2	288.0	19.45
31.6	1.38405	1.1366	328.4	288.9	19.52
31.7	1.38423	1.1371	329.6	289.9	19.59
31.8	1.38441	1.1376	330.8	290.8	19.66
31.9	1.38459	1.1381	332.1	291.8	19.74
32.0	1.38478	1.1386	333.3	292.7	19.81
32.1	1.38496	1.1391	334.5	293.7	19.88
32.2	1.38514	1.1396	335.7	294.6	19.95
32.3	1.38532	1.1401	337.0	295.6	20.03
32.4	1.38551	1.1406	338.2	296.5	20.10
32.5	1.38569	1.1411	339.4	297.5	20.17
32.6	1.38587	1.1416	340.7	298.4	20.25
32.7	1.38606	1.1421	341.9	299.4	20.32
32.8	1.38624	1.1426	343.1	300.3	20.39
32.9	1.38643	1.1431	344.4	301.3	20.47
33.0	1.38661	1.1436	345.6	302.2	20.54
33.1	1.38679	1.1441	346.8	303.2	20.61
33.2	1.38698	1.1446	348.1	304.1	20.69
33.3	1.38716	1.1451	349.3	305.0	20.76
33.4	1.38735	1.1456	350.6	306.0	20.84
33.5	1.38753	1.1461	351.8	306.9	20.91
33.6	1.38772	1.1466	353.0	307.9	20.98
33.7	1.38790	1.1471	354.3	308.8	21.06
33.8	1.38809	1.1476	355.5	309.8	21.13
33.9	1.38827	1.1481	356.8	310.7	21.20
34.0	1.38846	1.1486	358.0	31.7	21.28
34.1	1.38864	1.1491	359.2	312.6	21.35
34.2	1.38883	1.1496	360.5	313.6	21.42
34.3	1.38902	1.1501	361.7	314.5	21.50
34.4	1.38920	1.1507	363.0	315.5	21.57
34.5	1.38939	1.1512	364.2	316.4	21.64
34.6	1.38958	1.1517	365.5	317.4	21.72
34.7	1.38976	1.1522	366.7	318.3	21.79
34.8	1.38995	1.1527	368.0	319.2	21.87
34.9	1.39014	1.1532	369.2	320.2	21.94
35.0	1.39032	1.1537	370.5	321.1	22.02
35.1	1.39051	1.1542	371.8	322.1	22.10
35.2	1.39070	1.1547	373.0	323.0	22.17
35.3	1.39088	1.1552	374.3	324.0	22.24
35.4	1.39107	1.1557	375.5	324.9	22.32
35.5	1.39126	1.1563	376.8	325.9	22.39

(continued)

Table 4.2 (continued)

Saccharose % (m/m).	Refractive index at 20 °C	Mass density at 20 °C	Sugars in g/L	Sugars in g/kg	ABV % vol at 20 °C
35.6	1.39145	1.1568	378.0	326.8	22.46
35.7	1.39164	1.1573	379.3	327.8	22.54
35.8	1.39182	1.1578	380.6	328.7	22.62
35.9	1.39201	1.1583	381.8	329.6	22.69
36.0	1.39220	1.1588	383.1	330.6	22.77
36.1	1.39239	1.1593	384.4	331.5	22.84
36.2	1.39258	1.1598	385.6	332.5	22.92
36.3	1.39277	1.1603	386.9	333.4	22.99
36.4	1.39296	1.1609	388.1	334.4	23.06
36.5	1.39314	1.1614	389.4	335.3	23.14
36.6	1.39333	1.1619	390.7	336.3	23.22
36.7	1.39352	1.1624	392.0	337.2	23.30
36.8	1.39371	1.1629	393.2	338.1	23.37
36.9	1.39390	1.1634	394.5	339.1	23.45
37.0	1.39409	1.1640	395.8	340.0	23.52
37.1	1.39428	1.1645	397.0	341.0	23.59
37.2	1.39447	1.1650	398.3	341.9	23.67
37.3	1.39466	1.1655	399.6	342.9	23.75
37.4	1.39485	1.1660	400.0	343.8	23.83
37.5	1.39504	1.1665	402.1	344.7	23.90
37.6	1.39524	1.1671	403.4	345.7	23.97
37.7	1.39543	1.1676	404.7	346.6	24.05
37.8	1.39562	1.1681	406.0	347.6	24.13
37.9	1.39581	1.1686	407.3	348.5	24.21
38.0	1.39600	1.1691	408.6	349.4	24.28
38.1	1.39619	1.1697	409.8	350.4	24.35
38.2	1.39638	1.1702	411.1	351.3	24.43
38.3	1.39658	1.1707	412.4	352.3	24.51
38.4	1.39677	1.1712	413.7	353.2	24.59
38.5	1.39696	1.1717	415.0	354.2	24.66
38.6	1.39715	1.1723	416.3	355.1	24.74
38.7	1.39734	1.1728	417.6	356.0	24.82
38.8	1.39754	1.1733	418.8	357.0	24.89
38.9	1.39773	1.1738	420.1	357.9	24.97
39.0	1.39792	1.1744	421.4	358.9	25.04
39.1	1.39812	1.1749	422.7	359.8	25.12
39.2	1.39831	1.1754	424.0	360.7	25.20
39.3	1.39850	1.1759	425.3	361.7	25.28
39.4	1.39870	1.1765	426.6	362.6	25.35
39.5	1.39889	1.1770	427.9	363.6	25.43
39.6	1.39908	1.1775	429.2	364.5	25.51
39.7	1.39928	1.1780	430.5	365.4	25.58
39.8	1.39947	1.1786	431.8	366.4	25.66

(continued)

Table 4.2 (continued)

Saccharose % (m/m).	Refractive index at 20 °C	Mass density at 20 °C	Sugars in g/L	Sugars in g/kg	ABV % vol at 20 °C
39.9	1.39967	1.1791	433.1	367.3	25.74
40.0	1.39986	1.1796	434.4	368.3	25.82
40.1	1.40006	1.1801	435.7	369.2	25.89
40.2	1.40025	1.1807	437.0	370.1	25.97
40.3	1.40044	1.1812	438.3	371.1	26.05
40.4	1.40064	1.1817	439.6	372.0	26.13
40.5	1.40083	1.1823	440.9	373.0	26.20
40.6	1.40103	1.1828	442.2	373.9	26.28
40.7	1.40123	1.1833	443.6	374.8	26.36
40.8	1.40142	1.1839	444.9	375.8	26.44
40.9	1.40162	1.1844	446.2	376.7	26.52
41.0	1.40181	1.1849	447.5	377.7	26.59
41.1	1.40201	1.1855	448.8	378.6	26.67
41.2	1.40221	1.1860	450.1	379.5	26.75
41.3	1.40240	1.1865	451.4	380.5	26.83
41.4	1.40260	1.1871	452.8	381.4	26.91
41.5	1.40280	1.1876	454.1	382.3	26.99
41.6	1.40299	1.1881	455.4	383.3	27.06
41.7	1.40319	1.1887	456.7	384.2	27.14
41.8	1.40339	1.1892	458.0	385.2	27.22
41.9	1.40358	1.1897	459.4	386.1	27.30
42.0	1.40378	1.1903	460.7	387.0	27.38
42.1	1.40398	1.1908	462.0	388.0	27.46
42.2	1.40418	1.1913	463.3	388.9	27.53
42.3	1.40437	1.1919	464.7	389.9	27.62
42.4	1.40457	1.1924	466.0	390.8	27.69
42.5	1.40477	1.1929	467.3	391.7	27.77
42.6	1.40497	1.1935	468.6	392.7	27.85
42.7	1.40517	1.1940	470.0	393.6	27.93
42.8	1.40537	1.1946	471.3	394.5	28.01
42.9	1.40557	1.1951	472.6	395.5	28.09
43.0	1.40576	1.1956	474.0	396.4	28.17
43.1	1.40596	1.1962	475.3	397.3	28.25
43.2	1.40616	1.1967	476.6	398.3	28.32
43.3	1.40636	1.1973	478.0	399.2	28.41
43.4	1.40656	1.1978	479.3	400.2	28.48
43.5	1.40676	1.1983	480.7	401.1	28.57
43.6	1.40696	1.1989	482.0	402.0	28.65
43.7	1.40716	1.1994	483.3	403.0	28.72
43.8	1.40736	1.2000	484.7	403.9	28.81
43.9	1.40756	1.2005	486.0	404.8	28.88
44.0	1.40776	1.2011	487.4	405.8	28.97
44.1	1.40796	1.2016	488.7	406.7	29.04

(continued)

Table 4.2 (continued)

Saccharose % (m/m).	Refractive index at 20 °C	Mass density at 20 °C	Sugars in g/L	Sugars in g/kg	ABV % vol at 20 °C
44.2	1.40817	1.2022	490.1	407.6	29.13
44.3	1.40837	1.2027	491.4	408.6	29.20
44.4	1.40857	1.2032	492.8	409.5	29.29
44.5	1.40877	1.2038	494.1	410.4	29.36
44.6	1.40897	1.2043	495.5	411.4	29.45
44.7	1.40917	1.2049	496.8	412.3	29.52
44.8	1.40937	1.2054	498.2	413.3	29.61
44.9	1.40958	1.2060	499.5	414.2	29.69
45.0	1.40978	1.2065	500.9	415.1	29.77
45.1	1.40998	1.2071	502.2	416.1	29.85
45.2	1.41018	1.2076	503.6	417.0	29.93
45.3	1.41039	1.2082	504.9	417.9	30.01
45.4	1.41059	1.2087	506.3	418.9	30.09
45.5	1.41079	1.2093	507.7	419.8	30.17
45.6	1.41099	1.2098	509.0	420.7	30.25
45.7	1.41120	1.2104	510.4	421.7	30.33
45.8	1.41140	1.2109	511.7	422.6	30.41
45.9	1.41160	1.2115	513.1	423.5	30.49
46.0	1.41181	1.2120	514.5	424.5	30.58
46.1	1.41201	1.2126	515.8	425.4	30.65
46.2	1.41222	1.2131	517.2	426.3	30.74
46.3	1.41242	1.2137	518.6	427.3	30.82
46.4	1.41262	1.2142	519.9	428.2	30.90
46.5	1.41283	1.2148	521.3	429.1	30.98
46.6	1.41303	1.2154	522.7	430.1	31.06
46.7	1.41324	1.2159	524.1	431.0	31.15
46.8	1.41344	1.2165	525.4	431.9	31.22
46.9	1.41365	1.2170	526.8	432.9	31.31
47.0	1.41385	1.2176	528.2	433.8	31.39
47.1	1.41406	1.2181	529.6	434.7	31.47
47.2	1.41427	1.2187	530.9	435.7	31.55
47.3	1.41447	1.2192	532.3	436.6	31.63
47.4	1.41468	1.2198	533.7	437.5	31.72
47.5	1.41488	1.2204	535.1	438.5	31.80
47.6	1.41509	1.2209	536.5	439.4	31.88
47.7	1.41530	1.2215	537.9	440.3	31.97
47.8	1.41550	1.2220	539.2	441.3	32.04
47.9	1.41571	1.2226	540.6	442.2	32.13
48.0	1.41592	1.2232	542.0	443.1	32.21
48.1	1.41612	1.2237	543.4	444.1	32.29
48.2	1.41633	1.2243	544.8	445.0	32.38
48.3	1.41654	1.2248	546.2	445.9	32.46
48.4	1.41674	1.2254	547.6	446.8	32.54

(continued)

Table 4.2 (continued)

Saccharose % (m/m).	Refractive index at 20 °C	Mass density at 20 °C	Sugars in g/L	Sugars in g/kg	ABV % vol at 20 °C
48.5	1.41695	1.2260	549.0	447.8	32.63
48.6	1.41716	1.2265	550.4	448.7	32.71
48.7	1.41737	1.2271	551.8	449.6	32.79
48.8	1.41758	1.2277	553.2	450.6	32.88
48.9	1.41779	1.2282	554.6	451.5	32.96
49.0	1.41799	1.2288	556.0	452.4	33.04
49.1	1.41820	1.2294	557.4	453.4	33.13
49.2	1.41841	1.2299	558.8	454.3	33.21
49.3	1.41862	1.2305	560.2	455.2	33.29
49.4	1.41883	1.2311	561.6	456.2	33.38
49.5	1.41904	1.2316	563.0	457.1	33.46
49.6	1.41925	1.2322	564.4	458.0	33.54
49.7	1.41946	1.2328	565.8	458.9	33.63
49.8	1.41967	1.2333	567.2	459.9	33.71
49.9	1.41988	1.2339	568.6	460.8	33.79
50.0	1.42009	1.2345	570.0	461.7	33.88
50.1	1.42030	1.2350	571.4	462.7	33.96
50.2	1.42051	1.2356	572.8	463.6	34.04
50.3	1.42072	1.2362	574.2	464.5	34.12
50.4	1.42093	1.2368	575.6	465.4	34.21
50.5	1.42114	1.2373	577.1	466.4	34.30
50.6	1.42135	1.2379	578.5	467.3	34.38
50.7	1.42156	1.2385	579.9	468.2	34.46
50.8	1.42177	1.2390	581.3	469.2	34.55
50.9	1.42199	1.2396	582.7	470.1	34.63
51.0	1.42220	1.2402	584.2	471.0	34.72
51.1	1.42241	1.2408	585.6	471.9	34.80
51.2	1.42262	1.2413	587.0	472.9	34.89
51.3	1.42283	1.2419	588.4	473.8	34.97
51.4	1.42305	1.2425	589.9	474.7	35.06
51.5	1.42326	1.2431	591.3	475.7	35.14
51.6	1.42347	1.2436	592.7	476.6	35.22
51.7	1.42368	1.2442	594.1	477.5	35.31
51.8	1.42390	1.2448	595.6	478.4	35.40
51.9	1.42411	1.2454	597.0	479.4	35.48
52.0	1.42432	1.2460	598.4	480.3	35.56
52.1	1.42454	1.2465	599.9	481.2	35.65
52.2	1.42475	1.2471	601.3	482.1	35.74
52.3	1.42496	1.2477	602.7	483.1	35.82
52.4	1.42518	1.2483	604.2	484.0	35.91
52.5	1.42539	1.2488	605.6	484.9	35.99
52.6	1.42561	1.2494	607.0	485.8	36.07
52.7	1.42582	1.2500	608.5	486.8	36.16

(continued)

Table 4.2 (continued)

Saccharose % (m/m).	Refractive index at 20 °C	Mass density at 20 °C	Sugars in g/L	Sugars in g/kg	ABV % vol at 20 °C
52.8	1.42604	1.2506	609.9	487.7	36.25
52.9	1.42625	1.2512	611.4	488.6	36.34
53.0	1.42647	1.2518	612.8	489.5	36.42
53.1	1.42668	1.2523	614.2	490.5	36.50
53.2	1.42690	1.2529	615.7	491.4	36.59
53.3	1.42711	1.2535	617.1	492.3	36.67
53.4	1.42733	1.2541	618.6	493.2	36.76
53.5	1.42754	1.2547	620.0	494.2	36.85
53.6	1.42776	1.2553	621.5	495.1	36.94
53.7	1.42798	1.2558	622.9	496.0	37.02
53.8	1.42819	1.2564	624.4	496.9	37.11
53.9	1.42841	1.2570	625.8	497.9	37.19
54.0	1.42863	1.2576	627.3	498.8	37.28
54.1	1.42884	1.2582	628.7	499.7	37.36
54.2	1.42906	1.2588	630.2	500.6	37.45
54.3	1.42928	1.2594	631.7	501.6	37.54
54.4	1.42949	1.2600	633.1	502.5	37.63
54.5	1.42971	1.2606	634.6	503.4	37.71
54.6	1.42993	1.2611	636.0	504.3	37.80
54.7	1.43015	1.2617	637.5	505.2	37.89
54.8	1.43036	1.2623	639.0	506.2	37.98
54.9	1.43058	1.2629	640.4	507.1	38.06
55.0	1.43080	1.2635	641.9	508.0	38.15
55.1	1.43102	1.2641	643.4	508.9	38.24
55.2	1.43124	1.2647	644.8	509.9	38.32
55.3	1.43146	1.2653	646.3	510.8	38.41
55.4	1.43168	1.2659	647.8	511.7	38.50
55.5	1.43189	1.2665	649.2	512.6	38.58
55.6	1.43211	1.2671	650.7	513.5	38.67
55.7	1.43233	1.2677	652.2	514.5	38.76
55.8	1.43255	1.2683	653.7	515.4	38.85
55.9	1.43277	1.2689	655.1	516.3	38.93
56.0	1.43299	1.2695	656.6	517.2	39.02
56.1	1.43321	1.2701	658.1	518.1	39.11
56.2	1.43343	1.2706	659.6	519.1	39.20
56.3	1.43365	1.2712	661.0	520.0	39.28
56.4	1.43387	1.2718	662.5	520.9	39.37
56.5	1.43410	1.2724	664.0	521.8	39.46
56.6	1.43432	1.2730	665.5	522.7	39.55
56.7	1.43454	1.2736	667.0	523.7	39.64
56.8	4.43476	1.2742	668.5	524.6	39.73
56.9	1.43498	1.2748	669.9	525.5	39.81
57.0	1.43520	1.2754	671.4	526.4	39.90

(continued)

Table 4.2 (continued)

Saccharose % (m/m).	Refractive index at 20 °C	Mass density at 20 °C	Sugars in g/L	Sugars in g/kg	ABV % vol at 20 °C
57.1	1.43542	1.2760	672.9	527.3	39.99
57.2	1.43565	1.2766	674.4	528.3	40.08
57.3	1.43587	1.2773	675.9	529.2	40.17
57.4	1.43609	1.2779	677.4	530.1	40.26
57.5	1.43631	1.2785	678.9	531.0	40.35
57.6	1.43653	1.2791	680.4	531.9	40.44
57.7	1.43676	1.2797	681.9	532.8	40.53
57.8	1.43698	1.2803	683.4	533.8	40.61
57.9	1.43720	1.2809	684.9	534.7	40.70
58.0	1.43743	1.2815	686.4	535.6	40.79
58.1	1.43765	1.2821	687.9	536.5	40.88
58.2	1.43787	1.2827	689.4	537.4	40.97
58.3	1.43810	1.2833	690.9	538.3	41.06
58.4	1.43832	1.2839	692.4	539.3	41.15
58.5	1.43855	1.2845	693.9	540.2	41.24
58.6	1.43877	1.2851	695.4	541.1	41.33
58.7	1.43899	1.2857	696.9	542.0	41.42
58.8	1.43922	1.2863	698.4	542.9	41.51
58.9	1.43944	1.2870	699.9	543.8	41.60
59.0	1.43967	1.2876	701.4	544.8	41.68
59.1	1.43989	1.2882	702.9	545.7	41.77
59.2	1.44012	1.2888	704.4	546.6	41.86
59.3	1.44035	1.2894	706.0	547.5	41.96
59.4	1.44057	1.2900	707.5	548.4	42.05
59.5	1.44080	1.2906	709.0	549.3	42.14
59.6	1.44102	1.2912	710.5	550.2	42.23
59.7	1.44125	1.2919	712.0	551.1	42.31
59.8	1.44148	1.2925	713.5	552.1	42.40
59.9	1.44170	1.2931	715.1	553.0	42.50
60.0	1.44193	1.2937	716.6	553.9	42.59
60.1	1.44216	1.2943	718.1	554.8	42.68
60.2	1.44238	1.2949	719.6	555.7	42.77
60.3	1.44261	1.2956	721.1	556.6	42.85
60.4	1.44284	1.2962	722.7	557.5	42.95
60.5	1.44306	1.2968	724.2	558.4	43.04
60.6	1.44329	1.2974	725.7	559.4	43.13
60.7	1.44352	1.2980	727.3	560.3	43.22
60.8	1.44375	1.2986	728.8	561.2	43.31
60.9	1.44398	1.2993	730.3	562.1	43.40
61.0	1.44420	1.2999	731.8	563.0	43.49
61.1	1.44443	1.3005	733.4	563.9	43.59
61.2	1.44466	1.3011	734.9	564.8	43.68
61.3	1.44489	1.3017	736.4	565.7	43.76

(continued)

Table 4.2 (continued)

Saccharose % (m/m).	Refractive index at 20 °C	Mass density at 20 °C	Sugars in g/L	Sugars in g/kg	ABV % vol at 20 °C
61.4	1.44512	1.3024	738.0	566.6	43.86
61.5	1.44535	1.3030	739.5	567.6	43.95
61.6	1.44558	1.3036	741.1	568.5	44.04
61.7	1.44581	1.3042	742.6	569.4	44.13
61.8	1.44604	1.3049	744.1	570.3	44.22
61.9	1.44627	1.3055	745.7	571.2	44.32
62.0	1.44650	1.3061	747.2	572.1	44.41
62.1	1.44673	1.3067	748.8	573.0	44.50
62.2	1.44696	1.3074	750.3	573.9	44.59
62.3	1.44719	1.3080	751.9	547.8	44.69
62.4	1.44742	1.3086	753.4	575.7	44.77
62.5	1.44765	1.3092	755.0	576.6	44.87
62.6	1.44788	1.3099	756.5	577.5	44.96
62.7	1.44811	1.3105	758.1	578.5	45.05
62.8	1.44834	1.3111	759.6	579.4	45.14
62.9	1.44858	1.3118	761.2	580.3	45.24
63.0	1.44881	1.3124	762.7	581.2	45.33
63.1	1.44904	1.3130	764.3	582.1	45.42
63.2	1.44927	1.3137	765.8	583.0	45.51
63.3	1.44950	1.3143	767.4	583.9	45.61
63.4	1.44974	1.3149	769.0	584.8	45.70
63.5	1.44997	1.3155	770.5	585.7	45.79
63.6	1.45020	1.3162	772.1	586.6	45.89
63.7	1.45043	1.3168	773.6	587.5	45.98
63.8	1.45067	1.3174	775.2	588.4	46.07
63.9	1.45090	1.3181	776.8	589.3	46.17
64.0	1.45113	1.3187	778.3	590.2	46.25
64.1	1.45137	1.3193	779.9	591.1	46.35
64.2	1.45160	1.3200	781.5	592.0	46.44
64.3	1.45184	1.3206	783.0	592.9	46.53
64.4	1.45207	1.3213	784.6	593.8	46.63
64.5	1.45230	1.3219	786.2	594.7	46.72
64.6	1.45254	1.3225	787.8	595.6	46.82
64.7	1.45277	1.3232	789.3	596.5	46.91
64.8	1.45301	1.3238	790.9	597.4	47.00
64.9	1.45324	1.3244	792.5	598.3	47.10
65.0	1.45348	1.3251	794.1	599.3	47.19
65.1	1.45371	1.3257	795.6	600.2	47.28
65.2	1.45395	1.3264	797.2	601.1	47.38
65.3	1.45418	1.3270	798.8	602.0	47.47
65.4	1.45442	1.3276	800.4	602.9	47.57
65.5	1.45466	1.3283	802.0	603.8	47.66
65.6	1.45489	1.3289	803.6	604.7	47.76

(continued)

Table 4.2 (continued)

Saccharose % (m/m).	Refractive index at 20 °C	Mass density at 20 °C	Sugars in g/L	Sugars in g/kg	ABV % vol at 20 °C
65.7	1.45513	1.3296	805.1	605.6	47.85
65.8	1.45537	1.3302	806.7	606.5	47.94
65.9	1.45560	1.3309	808.3	607.4	48.04
66.0	1.45584	1.3315	809.9	608.3	48.13
66.1	1.45608	1.3322	811.5	609.2	48.23
66.2	1.45631	1.3328	813.1	610.1	48.32
66.3	1.45655	1.3334	814.7	611.0	48.42
66.4	1.45679	1.3341	816.3	611.9	48.51
66.5	1.45703	1.3347	817.9	612.8	48.61
66.6	1.45726	1.3354	819.5	613.7	48.70
66.7	1.45750	1.3360	821.1	614.6	48.80
66.8	1.45774	1.3367	822.7	615.5	48.89
66.9	1.45798	1.3373	824.3	616.3	48.99
67.0	1.45822	1.3380	825.9	617.2	49.08
67.1	1.45846	1.3386	827.5	618.1	49.18
67.2	1.45870	1.3393	829.1	619.0	49.27
67.3	1.45893	1.3399	830.7	619.9	49.37
67.4	1.45917	1.3406	832.3	620.8	49.46
67.5	1.45941	1.3412	833.9	621.7	49.56
67.6	1.45965	1.3419	835.5	622.6	49.65
67.7	1.45989	1.3425	837.1	623.5	49.75
67.8	1.46013	1.3432	838.7	624.4	49.84
67.9	1.46037	1.3438	840.3	625.3	49.94
68.0	1.46061	1.3445	841.9	626.2	50.03
68.1	1.46085	1.3451	843.6	627.1	50.14
68.2	1.46109	1.3458	845.2	628.0	50.23
68.3	1.46134	1.3464	846.8	628.9	50.33
68.4	1.46158	1.3471	848.4	629.8	50.42
68.5	1.46182	1.3478	850.0	630.7	50.52
68.6	1.46206	1.3484	851.6	631.6	50.61
68.7	1.46230	1.3491	853.3	632.5	50.71
68.8	1.46254	1.3497	854.9	633.4	50.81
68.9	1.46278	1.3504	856.5	634.3	50.90
69.0	1.46303	1.3510	858.1	635.2	51.00
69.1	1.46327	1.3517	859.8	636.1	51.10
69.2	1.46351	1.3524	861.4	636.9	51.19
69.3	1.46375	1.3530	863.0	637.8	51.29
69.4	1.46400	1.3537	864.7	638.7	51.39
69.5	1.46424	1.3543	866.3	639.6	51.48
69.6	1.46448	1.3550	867.9	640.5	51.58
69.7	1.46473	1.3557	869.5	641.4	51.67
69.8	1.46497	1.3563	871.2	642.3	51.78
69.9	1.46521	1.3570	872.8	643.2	51.87

(continued)

Table 4.2 (continued)

Saccharose % (m/m).	Refractive index at 20 °C	Mass density at 20 °C	Sugars in g/L	Sugars in g/kg	ABV % vol at 20 °C
70.0	1.46546	1.3576	874.5	644.1	51.97
70.1	1.46570	1.3583	876.1	645.0	52.07
70.2	1.46594	1.3590	877.7	645.9	52.16
70.3	1.46619	1.3596	879.4	646.8	52.26
70.4	1.46643	1.3603	881.0	647.7	52.36
70.5	1.46668	1.3610	882.7	648.5	52.46
70.6	1.46692	1.3616	884.3	649.4	52.55
70.7	1.46717	1.3623	886.0	650.3	52.65
70.8	1.46741	1.3630	887.6	651.2	52.75
70.9	1.46766	1.3636	889.3	652.1	52.85
71.0	1.46790	1.3643	890.9	653.0	52.95
71.1	1.46815	1.3650	892.6	653.9	53.05
71.2	1.46840	1.3656	894.2	654.8	53.14
71.3	1.46864	1.3663	895.9	655.7	53.24
71.4	1.46889	1.3670	897.5	656.6	53.34
71.5	1.46913	1.3676	899.2	657.5	53.44
71.6	1.46938	1.3683	900.2	658.3	53.53
71.7	1.46963	1.3690	902.5	659.2	53.64
71.8	1.46987	1.3696	904.1	660.1	53.73
71.9	1.47012	1.3703	905.8	661.0	53.83
72.0	1.47037	1.3710	907.5	661.9	53.93
72.1	1.47062	1.3717	909.1	662.8	54.03
72.2	1.47086	1.3723	910.8	663.7	54.13
72.3	1.47111	1.3730	912.5	664.6	54.23
72.4	1.47136	1.3737	914.1	665.5	54.32
72.5	1.47161	1.3743	915.8	666.3	54.43
72.6	1.47186	1.3750	917.5	667.2	54.53
72.7	1.47210	1.3757	919.1	668.1	54.62
72.8	1.47235	1.3764	920.8	669.0	54.72
72.9	1.47260	1.3770	922.5	669.9	54.82
73.0	1.47285	1.3777	924.2	670.8	54.93
73.1	1.47310	1.3784	925.8	671.7	55.02
73.2	1.47335	1.3791	927.5	672.6	55.12
73.3	1.47360	1.3797	929.2	673.5	55.22
73.4	1.47385	1.3804	930.9	674.3	55.32
73.5	1.47410	1.3811	932.6	675.2	55.42
73.6	1.47435	1.3818	934.3	676.1	55.53
73.7	1.47460	1.3825	935.9	677.0	55.62
73.8	1.47485	1.3831	937.6	677.9	55.72
73.9	1.47510	1.3838	939.3	678.8	55.82
74.0	1.47535	1.3845	941.0	679.7	55.92
74.1	1.47560	1.3852	942.7	680.6	56.02
74.2	1.47585	1.3859	944.4	681.4	56.13

(continued)

Table 4.2 (continued)

Saccharose % (m/m).	Refractive index at 20 °C	Mass density at 20 °C	Sugars in g/L	Sugars in g/kg	ABV % vol at 20 °C
74.3	1.47610	1.3865	946.1	682.3	56.23
74.4	1.47635	1.3872	947.8	683.2	56.33
74.5	1.47661	1.3879	949.5	684.1	56.43
74.6	1.47686	1.3886	951.2	685.0	56.53
74.7	1.47711	1.3893	952.9	685.9	56.63
74.8	1.47736	1.3899	954.6	686.8	56.73
74.9	1.47761	1.3906	956.3	687.7	56.83

Table 4.3 The sugar concentration in rectified concentrated must in grams per liter and grams per kilogram. Determined by means of a refractometer graduated either in percentage by mass of sucrose at 20 °C or in refractive index at 20 °C

Saccharose % (m/m)	Refractive index at 20 °C	Mass density at 20 °C	Sugars in g/L	Sugars in g/kg	ABV % vol at 20 °C
55.0	1.43079	1.2635	707.8	560.2	42.06
55.1	1.43102	1.2639	709.4	561.3	42.16
55.2	1.43124	1.2645	711	562.3	42.25
55.3	1.43146	1.2652	712.7	563.3	42.36
55.4	1.43168	1.2659	714.4	564.3	42.46
55.5	1.43189	1.2665	716.1	565.4	42.56
55.6	1.43211	1.2672	717.8	566.4	42.66
55.7	1.43233	1.2679	719.5	567.5	42.76
55.8	1.43255	1.2685	721.1	568.5	42.85
55.9	1.43277	1.2692	722.8	569.5	42.96
56.0	1.43298	1.2699	724.5	570.5	43.06
56.1	1.43321	1.2703	726.1	571.6	43.15
56.2	1.43343	1.2708	727.7	572.6	43.2
56.3	1.43365	1.2713	729.3	573.7	43.34
56.4	1.43387	1.2718	730.9	574.7	43.44
56.5	1.43409	1.2724	732.6	575.8	43.54
56.6	1.43431	1.2731	734.3	576.8	43.64
56.7	1.43454	1.2738	736	577.8	43.74
56.8	1.43476	1.2744	737.6	578.8	43.84
56.9	1.43498	1.2751	739.4	579.9	43.94
57.0	1.43519	1.2758	741.1	580.9	44.04
57.1	1.43542	1.2763	742.8	582	44.14
57.2	1.43564	1.2768	744.4	583	44.24
57.3	1.43586	1.2773	745.9	584	44.33
57.4	1.43609	1.2778	747.6	585.1	44.43
57.5	1.43631	1.2784	749.3	586.1	44.53
57.6	1.43653	1.2791	751	587.1	44.63
57.7	1.43675	1.2798	752.7	588.1	44.73
57.8	1.43698	1.2804	754.4	589.2	44.83

(continued)

Table 4.3 (continued)

Saccharose % (m/m)	Refractive index at 20 °C	Mass density at 20 °C	Sugars in g/L	Sugars in g/kg	ABV % vol at 20 °C
57.9	1.4372	1.281	756.1	590.2	44.94
58.0	1.43741	1.2818	757.8	591.2	45.04
58.1	1.43764	1.2822	759.5	592.3	45.14
58.2	1.43784	1.2827	761.1	593.4	45.23
58.3	1.43909	1.2832	762.6	594.3	45.32
58.4	1.43832	1.2837	764.3	595.4	45.42
58.5	1.43854	1.2843	766	596.4	45.52
58.6	1.43877	1.285	767.8	597.5	45.63
58.7	1.43899	1.2857	769.5	598.5	45.73
58.8	1.43922	1.2863	771.1	599.5	45.83
58.9	1.43944	1.2869	772.9	600.6	45.93
59.0	1.43966	1.2876	774.6	601.6	46.03
59.1	1.43988	1.2882	776.3	602.6	46.14
59.2	1.44011	1.2889	778.1	603.7	46.24
59.3	1.44034	1.2896	779.8	604.7	46.34
59.4	1.44057	1.2902	781.6	605.8	46.45
59.5	1.44079	1.2909	783.3	606.8	46.55
59.6	1.44102	1.2916	785.2	607.9	46.66
59.7	1.44124	1.2921	786.8	608.9	46.76
59.8	1.44147	1.2926	788.4	609.9	46.85
59.9	1.44169	1.2931	790	610.9	46.95
60.0	1.44192	1.2936	791.7	612	47.05
60.1	1.44215	1.2942	793.3	613	47.15
60.2	1.44238	1.2949	795.2	614.1	47.26
60.3	1.4426	1.2956	796.9	615.1	47.36
60.4	1.44283	1.2962	798.6	616.1	47.46
60.5	1.44305	1.2969	800.5	617.2	47.57
60.6	1.44328	1.2976	802.2	618.2	47.67
60.7	1.44351	1.2981	803.9	619.3	47.78
60.8	1.44374	1.2986	805.5	620.3	47.87
60.9	1.44397	1.2991	807.1	621.3	47.97
61.0	1.44419	1.2996	808.7	622.3	48.06
61.1	1.44442	1.3002	810.5	623.4	48.17
61.2	1.44465	1.3009	812.3	624.4	48.27
61.3	1.44488	1.3016	814.2	625.5	48.39
61.4	1.44511	1.3022	815.8	626.5	48.48
61.5	1.44534	1.3029	817.7	627.6	48.6
61.6	1.44557	1.3036	819.4	628.6	48.7
61.7	1.4458	1.3042	821.3	629.7	48.81
61.8	1.44603	1.3049	823	630.7	48.91
61.9	1.44626	1.3056	824.8	631.7	49.02
62.0	1.44648	1.3062	826.6	632.8	49.12
62.1	1.44672	1.3068	828.3	633.8	49.23

(continued)

Table 4.3 (continued)

Saccharose % (m/m)	Refractive index at 20 °C	Mass density at 20 °C	Sugars in g/L	Sugars in g/kg	ABV % vol at 20 °C
62.2	1.44695	1.3075	830	634.8	49.33
62.3	1.44718	1.308	831.8	635.9	49.43
62.4	1.44741	1.3085	833.4	636.9	49.53
62.5	1.44764	1.309	835.1	638	49.63
62.6	1.44787	1.3095	836.8	639	49.73
62.7	1.4481	1.3101	8.85	640	49.83
62.8	1.44833	1.3108	840.2	641	49.93
62.9	1.44856	1.3115	842.1	642.1	50.05
63.0	1.44879	1.3121	843.8	643.1	50.15
63.1	1.44902	1.3128	845.7	644.2	50.26
63.2	1.44926	1.3135	847.5	645.2	50.37
63.3	1.44949	1.3141	849.3	646.3	50.47
63.4	1.44972	1.3148	851.1	647.3	50.58
63.5	1.44995	1.3155	853	648.4	50.69
63.6	1.45019	1.3161	854.7	649.4	50.79
63.7	1.45042	1.3168	856.5	650.4	50.9
63.8	1.45065	1.3175	858.4	651.5	51.01
63.9	1.45088	1.318	860	652.5	51.11
64.0	1.45112	1.3185	861.6	653.5	51.2
64.1	1.45135	1.319	863.4	654.6	51.31
64.2	1.45158	1.3195	865.1	655.6	51.41
64.3	1.45181	1.3201	866.9	656.7	51.52
64.4	1.45205	1.3208	868.7	657.7	51.63
64.5	1.45228	1.3215	870.6	658.8	51.74
64.6	1.45252	1.3221	872.3	659.8	51.84
64.7	1.45275	1.3228	874.1	660.8	51.95
64.8	1.45299	1.3235	876	661.9	52.06
64.9	1.45322	1.3241	877.8	662.9	52.17
65.0	1.45347	1.3248	879.7	664	52.28
65.1	1.45369	1.3255	881.5	665	52.39
65.2	1.45393	1.3261	883.2	666	52.49
65.3	1.45416	1.3268	885	667	52.6
65.4	1.4544	1.3275	886.9	668.1	52.71
65.5	1.45463	1.3281	888.8	669.2	52.82
65.6	1.45487	1.3288	890.6	670.2	52.93
65.7	1.4551	1.3295	892.4	671.2	53.04
65.8	1.45534	1.3301	894.2	672.3	53.14
65.9	1.45557	1.3308	896	673.3	53.25
66.0	1.45583	1.3315	898	674.4	53.37
66.1	1.45605	1.332	899.6	675.4	53.46
66.2	1.45629	1.3325	901.3	676.4	53.56
66.3	1.45652	1.333	903.1	677.5	53.67
66.4	1.45676	1.3335	904.8	678.5	53.77

(continued)

Table 4.3 (continued)

Saccharose % (m/m)	Refractive index at 20 °C	Mass density at 20 °C	Sugars in g/L	Sugars in g/kg	ABV % vol at 20 °C
66.5	1.457	1.3341	906.7	679.6	53.89
66.6	1.45724	1.3348	908.5	680.6	53.99
66.7	1.45747	1.3355	910.4	681.7	54.11
66.8	1.45771	1.3361	912.2	682.7	54.21
66.9	1.45795	1.3367	913.9	983.7	54.31
67.0	1.4582	1.3374	915.9	984.8	54.43
67.1	1.45843	1.338	917.6	685.8	54.53
67.2	1.45867	1.3387	919.6	686.9	54.65
67.3	1.4589	1.3395	921.4	687.9	54.76
67.4	1.45914	1.34	923.1	688.9	54.86
67.5	1.45938	1.3407	925.1	690	54.98
67.6	1.45962	1.3415	927	691	55.09
67.7	1.45986	1.342	928.8	692.1	55.2
67.8	1.4601	1.3427	930.6	693.1	55.31
67.9	1.46034	1.3434	932.6	694.2	55.42
68.0	1.4606	1.344	934.4	695.2	55.53
68.1	1.46082	1.3447	936.2	696.2	55.64
68.2	1.46106	1.3454	938	697.2	55.75
68.3	1.4613	1.346	939.9	698.3	55.86
68.4	1.46154	1.3466	941.8	699.4	55.97
68.5	1.46178	1.3473	943.7	700.4	56.08
68.6	1.46202	1.3479	945.4	701.4	56.19
68.7	1.46226	1.3486	947.4	702.5	56.3
68.8	1.46251	1.3493	949.2	703.5	56.41
68.9	1.46275	1.3499	951.1	704.6	56.52
69.0	1.46301	1.3506	953	705.6	56.64
69.1	1.46323	1.3513	954.8	706.6	56.74
69.2	1.43347	1.3519	956.7	707.7	56.86
69.3	1.46371	1.3526	958.6	708.7	56.97
69.4	1.46396	1.3533	960.6	709.8	57.09
69.5	1.4642	1.3539	962.4	710.8	57.2
69.6	1.46444	1.3546	964.3	711.9	57.31
69.7	1.46468	1.3553	966.2	712.9	57.42
69.8	1.46493	1.356	968.2	714	57.54
69.9	1.46517	1.3566	970	715	57.65
70.0	1.46544	1.3573	971.8	716	57.75
70.1	1.46595	1.3579	973.8	717.1	57.87
70.2	1.4659	1.3586	975.6	718.1	57.98
70.3	1.46614	1.3593	977.6	719.2	58.1
70.4	1.46639	1.3599	979.4	720.2	58.21
70.5	1.46663	1.3606	981.3	721.2	58.32
70.6	1.46688	1.3613	983.3	722.3	58.44
70.7	1.46712	1.3619	985.2	723.4	58.55

(continued)

Table 4.3 (continued)

Saccharose % (m/m)	Refractive index at 20 °C	Mass density at 20 °C	Sugars in g/L	Sugars in g/kg	ABV % vol at 20 °C
70.8	1.46737	1.3626	987.1	724.4	58.66
70.9	1.46761	1.3633	988.9	725.4	58.77
71.0	1.46789	1.3639	990.9	726.5	58.89
71.1	1.4681	1.3646	992.8	727.5	59
71.2	1.46835	1.3653	994.8	728.6	59.12
71.3	1.46859	1.3659	996.6	729.6	59.23
71.4	1.46884	1.3665	998.5	730.7	59.34
71.5	1.46908	1.3672	1000.4	731.7	59.45
71.6	1.46933	1.3678	1002.2	732.7	59.56
71.7	1.46957	1.3685	1004.2	733.8	59.68
71.8	1.46982	1.3692	1006.1	734.8	59.79
71.9	1.74007	1.3698	1008	735.9	59.91
72.0	1.47036	1.3705	1009.9	736.9	60.02
72.1	1.47056	1.3712	1012	738	60.14
72.2	1.47081	1.3718	1013.8	739	60.25
72.3	1.47106	1.3725	1015.7	740	60.36
72.4	1.47131	1.3732	1017.7	741.1	60.48
72.5	1.47155	1.3738	1019.5	742.1	60.59
72.6	1.4718	1.3745	1021.5	743.2	60.71
72.7	1.47205	1.3752	1023.4	744.2	60.82
72.8	1.4723	1.3758	1025.4	745.3	60.94
72.9	1.47254	1.3765	1027.3	746.3	61.05
73.0	1.47284	1.3772	1029.3	747.4	61.17
73.1	1.47304	1.3778	1031.2	748.4	61.28
73.2	1.47329	1.3785	1033.2	749.5	61.4
73.3	1.47354	1.3792	1035.1	750.5	61.52
73.4	1.47379	1.3798	1037.1	751.6	61.63
73.5	1.47404	1.3805	1039	752.6	61.75
73.6	1.47429	1.3812	1040.9	753.6	61.86
73.7	1.47454	1.3818	1042.8	757.7	61.97
73.8	1.47479	1.3825	1044.8	755.7	62.09
73.9	1.47504	1.3832	1046.8	756.8	62.21
74.0	1.47534	1.3838	1048.6	757.8	62.32
74.1	1.47554	1.3845	1050.7	758.9	62.44
74.2	1.47579	1.3852	1052.6	759.9	62.56
74.3	1.47604	1.3858	1054.6	761	62.67
74.4	1.47629	1.3865	1056.5	762	62.79
74.5	1.47654	1.3871	1058.5	763.1	62.91
74.6	1.47679	1.3878	1060.4	764.1	63.02
74.7	1.47704	1.3885	1062.3	765.1	63.13
74.8	1.4773	1.3892	1064.4	766.2	63.26
74.9	1.47755	1.3898	1066.3	767.2	63.37
75.0	1.47785	1.3905	1068.3	768.3	63.49

If the measurement was made on diluted rectified concentrated must, multiply the result by the dilution factor.

Refractive Index of Must, Concentrated Must, and Rectified Concentrated Must

Find the percentage by mass of sucrose at 20 °C in Table 4.2 and read from the same row the refractive index at 20 °C. This index is expressed to four decimal places.

4.1.2 Sugars in Must with the Use of Baume Hydrometer

The most common method of measuring sugars in grape musts is by using hydrometers. This is a very easy method, fast, and very accurate.

Principle

This method is based on densitometry. More specifically, sugars make up most of the components that are dissolved in the wort (12–30% instead of 5–6% of the remaining ingredients). This has the consequence that the change in its specific gravity depends on its sugar content. Baume hydrometers are calibrated in degrees Baume (Be) and are calibrated at 20 °C (previously they were calibrated at 15 °C). This means that when measurements are made at temperatures greater or less than 20 °C they must be reduced to 20 °C.

Apparatus

1. **Baume hydrometer**
2. **Thermometer**
3. **Volumetric cylinder 250 mL**
4. **Grape must**

Preparation of the Grape Must Sample

The grape must we are going to measure should be free from suspensions. If we deem it necessary, we can filter the measured must by a lab filter or dry gauze.

Procedure

The hydrometer must be clean and dry before each use. 200 mL of grape must is added inside our volumetric cylinder. This process should be done slowly so that the must does not have foam or bubbles that can affect our measurement. We then place our volumetric cylinder vertically on our workbench and dip the hydrometer into the must. This will oscillate until it balances. When it is balanced, we push the hydrometer by finger by one degree and after it returns to the same position, we note the indication. The must is very likely to make a slight curve where it contacts the part of the hydrometer we use to read. We must bring an imaginary straight line to the base of the curve in order to measure the reading with great precision. Immediately after the reading, we take the temperature of the must and reduce the reading we have measured to 20 °C.

Calculation

After we have noted the reading of the hydrometer and the thermometer, we use the following formula in order to reduce the reading of the hydrometer to 20 °C.

$$Be_{(R)} = Be_{(T)} + (T - 20) \times 0.05$$

$Be_{(R)}$ = Real Baume
$Be_{(T)}$ = Baume measured my hydrometer
T = Temperature of the must

Sugar Concentration in Must and Concentrated Must

After finding the actual Baume grade of the must with the help of the Table 4.4, we can determine the sugar content, which is expressed in grams per liter.

Table 4.4 Correspondence of Baume degrees to density and sugars concentration

Baume.	Density	Sugars ‰	Baume	Density	Sugars ‰
0.0	1.00000	–	17.6	1.13890	334.2
0.1	1.00070	–	17.7	1.13980	336.3
0.2	1.00140	–	17.8	1.14070	338.5
0.3	1.00210	–	17.9	1.14160	340.7
0.4	1.00280	–	18.0	1.14250	342.8
0.5	1.00350	–	18.1	1.14342	345.0
0.6	1.00420	–	18.2	1.14432	347.2
0.7	1.00490	–	18.3	1.14523	349.4
0.8	1.00560	–	18.4	1.14614	351.6
0.9	1.00630	–	18.5	1.14705	353.8
1.0	1.00700	–	18.6	1.14796	356.0
1.1	1.00770	–	18.7	1.14887	358.2
1.2	1.00840	–	18.8	1.14978	360.4
1.3	1.00910	–	18.9	1.15069	362.6
1.4	1.00980	–	19.0	1.15160	364.8
1.5	1.01050	–	19.1	1.15253	367.0
1.6	1.01120	–	19.2	1.15346	369.2
1.7	1.01190	–	19.3	1.15439	371.4
1.8	1.01260	–	19.4	1.15532	373.7
1.9	1.01330	–	19.5	1.15625	375.9
2.0	1.01410	–	19.6	1.15718	378.1
2.1	1.01481	–	19.7	1.15811	380.3
2.2	1.01552	–	19.8	1.15904	382.6
2.3	1.01623	–	19.9	1.15997	384.8

(continued)

Table 4.4 (continued)

Baume.	Density	Sugars ‰	Baume	Density	Sugars ‰
2.4	1.01694	–	20.0	1.16090	387.1
2.5	1.01765	–	20.1	1.16184	389.3
2.6	1.01836	–	20.2	1.16278	391.6
2.7	1.01907	–	20.3	1.16372	393.8
2.8	1.01978	–	20.4	1.16466	396.1
2.9	1.02049	–	20.5	1.16560	398.3
3.0	1.02120	–	20.6	1.16654	400.6
3.1	1.02193	–	20.7	1.16746	402.9
3.2	1.02266	–	20.8	1.16842	405.2
3.3	1.02339	–	20.9	1.16939	407.4
3.4	1.02412	–	21.0	1.17030	409.7
3.5	1.02485	–	21.1	1.17126	412.0
3.6	1.02558	–	21.2	1.17222	414.0
3.7	1.02631	–	21.3	1.17318	416.6
3.8	1.02704	–	21.4	1.17414	418.9
3.9	1.02777	–	21.5	1.17610	421.2
4.0	1.02850	–	21.6	1.17606	423.5
4.1	1.02924	–	21.7	1.17702	425.8
4.2	1.02998	–	21.8	1.17798	428.1
4.3	1.03072	–	21.9	1.17894	430.5
4.4	1.03146	–	22.0	1.17990	432.7
4.5	1.03220	–	22.1	1.18087	435.1
4.6	1.03294	–	22.2	1.18184	437.4
4.7	1.03368	–	22.3	1.18281	439.7
4.8	1.03442	–	22.4	1.18378	442.1
4.9	1.03516	63.0	22.5	1.18475	444.4
5.0	1.03590	66.0	22.6	1.18572	446.7
5.1	1.03665	69.0	22.7	1.18669	449.1
5.2	1.03740	71.0	22.8	1.18766	451.4
5.3	1.03815	72.0	22.9	1.18863	453.8
5.4	1.03890	73.0	23.0	1.18960	456.1
5.5	1.03965	74.0	23.1	1.19057	458.5
5.6	1.04040	76.0	23.2	1.19158	460.9
5.7	1.04115	80.0	23.3	1.19257	463.2
5.8	1.04190	82.0	23.4	1.19356	465.6
5.9	1.04265	84.0	23.5	1.19455	468.0
6.0	1.04340	87.0	23.6	1.19554	470.4
6.1	1.04416	90.0	23.7	1.19653	472.8
6.2	1.04492	91.0	23.8	1.19752	475.1
6.3	1.04568	92.0	23.9	1.19851	477.5
6.4	1.04664	95.0	24.0	1.19950	479.9
6.5	1.04720	96.0	24.1	1.20050	482.3
6.6	1.04796	98.0	24.2	1.20150	484.7
6.7	1.04878	100.0	24.3	1.20250	487.1

(continued)

Table 4.4 (continued)

Baume.	Density	Sugars ‰	Baume	Density	Sugars ‰
6.8	1.04948	102.0	24.4	1.20350	489.6
6.9	1.05034	103.0	24.5	1.20450	492.0
7.0	1.05100	106.0	24.6	1.20550	494.4
7.1	1.05177	108.0	24.7	1.20650	496.8
7.2	1.05254	111.0	24.8	1.20750	499.2
7.3	1.05331	113.0	24.9	1.20580	501.7
7.4	1.05408	114.0	25.0	1.20950	504.1
7.5	1.05485	116.0	25.1	1.21052	506.6
7.6	1.05562	119.0	25.2	1.21154	509.0
7.7	1.05639	120.0	25.3	1.21256	511.5
7.8	1.05716	122.0	25.4	1.21358	513.9
7.9	1.05793	124.0	25.5	1.21460	516.4
8.0	1.05870	127.0	25.6	1.21562	518.8
8.1	1.05948	130.0	25.7	1.21644	521.3
8.2	1.06026	132.0	25.8	1.21766	523.8
8.3	1.06104	134.0	25.9	1.21868	526.2
8.4	1.06182	135.0	26.0	1.21970	528.7
8.5	1.06260	138.0	26.1	1.22074	531.2
8.6	1.06338	140.0	26.2	1.22178	533.7
8.7	1.06416	143.0	26.3	1.22282	536.2
8.8	1.06494	145.0	26.4	1.22386	538.7
8.9	1.06572	146.0	26.5	1.22490	541.2
9.0	1.06650	151.0	26.6	1.22594	543.7
9.1	1.06729	154.0	26.7	1.22698	546.2
9.2	1.06808	157.0	26.8	1.22802	548.7
9.3	1.06884	159.0	26.9	1.22906	551.2
9.4	1.06966	161.0	27.0	1.23010	553.7
9.5	1.07045	164.0	27.1	1.23116	556.2
9.6	1.07124	165.0	27.2	1.23222	558.8
9.7	1.07203	168.0	27.3	1.23328	561.2
9.8	1.07288	171.0	27.4	1.23434	563.9
9.9	1.07361	174.0	27.5	1.23540	566.4
10.0	1.07450	179.1	27.6	1.23046	568.9
10.1	1.07530	181.0	27.7	1.23752	571.5
10.2	1.07610	183	27.8	1.23958	574.0
10.3	1.07690	184.9	27.9	1.23964	576.6
10.4	1.07770	186.8	28.0	1.24070	579.2
10.5	1.07850	188.8	28.1	1.24178	581.7
10.6	1.07930	190.7	28.2	1.24286	584.3
10.7	1.08010	192.7	28.3	1.24394	586.9
10.8	1.08080	194.8	28.4	1.24502	589.5
10.9	1.08170	196.5	28.5	1.24610	592.1
11.0	1.08250	198.5	28.6	1.24718	594.6
11.1	1.08332	200.5	28.7	1.24726	597.2

(continued)

Table 4.4 (continued)

Baume.	Density	Sugars ‰	Baume	Density	Sugars ‰
11.2	1.08414	202.4	28.8	1.24934	599.8
11.3	1.08496	204.4	28.9	1.25042	602.4
11.4	1.08578	206.3	29.0	1.25150	605.1
11.5	1.08660	208.3	29.1	1.25259	607.7
11.6	1.87420	210.3	29.2	1.25368	610.3
11.7	1.08824	212.3	29.3	1.25477	612.9
11.8	4.08906	214.2	29.4	1.25586	615.5
11.9	1.08998	216.2	29.5	1.25695	618.2
12.0	1.09070	218.2	29.6	1.25804	620.8
12.1	1.09153	202.2	29.7	1.25913	623.4
12.2	1.09236	222.2	29.8	1.26022	626.1
12.3	1.09319	224.2	29.9	1.26131	628.7
12.4	1.09402	226.1	30.0	1.26240	631.4
12.5	1.09485	228.1	30.1	1.26352	634.0
12.6	1.09568	230.1	30.2	1.26464	636.7
12.7	1.09651	232.2	30.3	1.26576	639.4
12.8	1.09734	234.2	30.4	1.26688	642.1
12.9	1.09815	236.2	30.5	1.26800	644.7
13.0	1.09900	238.2	30.6	1.26912	647.4
13.1	1.09984	240.2	30.7	1.27024	650.1
13.2	1.10068	242.2	30.8	1.27136	652.8
13.3	1.10152	244.2	30.9	1.27248	655.5
13.4	1.10238	246.3	31.0	1.27360	658.2
13.5	1.10320	248.3	31.1	1.27473	660.9
13.6	1.10404	250.3	31.2	1.27586	663.6
13.7	1.10488	252.4	31.3	1.27699	666.3
13.8	1.10572	254.4	31.4	1.27812	669.0
13.9	1.10656	256.4	31.5	1.27925	671.8
14.0	1.10740	258.5	31.6	1.28038	674.5
14.1	1.10826	260.5	31.7	1.28151	677.2
14.2	1.10912	262.6	31.8	1.28264	680.0
14.3	1.10998	264.6	31.9	1.28377	682.7
14.4	1.11084	266.7	32.0	1.28490	685.5
14.5	1.11700	268.7	32.1	1.28605	688.3
14.6	1.11296	270.8	32.2	1.28720	691.0
14.7	1.11342	272.9	32.3	1.28835	693.8
14.8	1.11428	274.9	32.4	1.28950	696.5
14.9	1.11514	277	32.5	1.29065	699.3
15.0	1.11600	279.1	32.6	1.29180	702.1
15.1	1.11687	281.2	32.7	1.29295	704.9
15.2	1.11774	283.2	32.8	1.29410	707.7
15.3	1.11861	285.3	32.9	1.29525	710.5
15.4	1.11948	287.4	33.0	1.29640	713.2
15.5	1.12035	289.5	33.1	1.29758	716.0

(continued)

Table 4.4 (continued)

Baume.	Density	Sugars ‰	Baume	Density	Sugars ‰
15.6	1.12122	291.6	33.2	1.29876	718.9
15.7	1.12209	293.7	33.3	1.29964	721.7
15.8	1.12296	295.8	33.4	1.30112	724.5
15.9	1.12383	297.9	33.5	1.30230	727.3
16.0	1.12470	300.0	33.6	1.30348	730.1
16.1	1.12558	302.1	33.7	1.30466	733.0
16.2	1.12646	304.2	33.8	1.30584	735.8
16.3	1.12734	306.3	33.9	1.30702	738.7
16.4	1.12822	308.5	34.0	1.30820	741.5
16.5	1.12910	310.6	34.1	1.30940	744.4
16.6	1.12998	312.7	34.2	1.31060	474.2
16.7	1.13086	314.8	34.3	1.31180	750.1
16.8	1.13174	317.0	34.4	1.31300	753.0
16.9	1.13262	319.1	34.5	1.31420	755.8
17.0	1.13350	321.3	34.6	1.31540	758.7
17.1	1.13440	323.4	34.7	1.31660	761.6
17.2	1.13530	325.5	34.8	1.31780	764.5
17.3	1.13620	327.7	34.9	1.31900	767.4
17.4	1.13710	329.9	35.0	1.32020	770.3
17.5	1.13800	332.0			

4.2 Ethyl Alcohol

The ethyl alcohol content of wines is expressed by the alcoholic strength which is equal to the number of liters of ethyl alcohol contained in 100 L of wine. Both of these volumes are measured at 20 °C. The alcoholic strength is denoted by a number followed by a % vol symbol (e.g., 13.5% vol). There are several methods of calculating the alcoholic strength, the method we have chosen to present is the method with simple distillation and measurement of the alcoholic strength with alcoholmeters.

Principle

This method consists in the simple distillation of the liquid after making it alkaline. The density of the spirit is measured with an alcoholmeter and the alcoholic strength is determined to the nearest one-tenth (0.1% vol).

Apparatus

- Distilling device
- Bunsen burner
- A volumetric flask with a capacity of 200 mL
- An alcoholmeter with controlled accuracy, graduated in % vol

- A volumetric cylinder
- A thermometer
- Amount of 200 mL of the sample to be examined
- 10 mL of [$Ca(OH)_2$] 10% [120 g CaO in 1 L of distilled water 60–70 °C]
- Pumice grains

Procedure

We measure 200 mL of wine with the volumetric flask and transfer them to the spherical kettle of the distilling device. At the same time, we measure the temperature of the wine. The volumetric flask is rinsed three times in succession with 5 mL each time of distilled water.

To neutralize the acidity of the wine in order to prevent the passage of volatile acids (acetate acid, carbonate acid, and sulfite acid) into the distillate, we add 10 mL of [$Ca(OH)_2$] 10% to the spherical kettle where the wine is with the distilled water we used to rinse the volumetric flask. Then we add some pumice grains to prevent foaming during boiling.

After securing spherical kettle distilling device, we gently heat the spherical kettle. The resulting distillate is collected in the same volumetric flask used to measure the sample. A small amount of distilled water is placed inside the volumetric flask so that the air arriving from the end of the cooler, which is loaded with alcohol vapors, is washed away from the water and leaves the alcohol it carries. When we have collected a little over 180 mL of distillate we stop the distillation. At this point we must be careful to remove the volumetric flask that collects the distillate before turning off the fire, otherwise there is a possibility that the distillate will be sucked up and a quantity of it will be transferred back to the kettle.

After completing the distillation, bring the temperature of the distillate as close as possible to the temperature our wine had before the distillation and fill the volumetric flask up to 200 mL with distilled water. We stir very well and transfer the hydro alcoholic solution to a volumetric cylinder at an angle of 45° where with the help of an alcoholmeter we get the reading of the alcoholic strength (*A*). At the same time, we measure the temperature of the distillate (*T*).

Calculation

The alcoholmeters we use now are calibrated at 20 °C. Since the distillate temperature measurement is exactly 20 °C, then we do not need to use correction tables to express the result. Otherwise we use Table 4.5 to correct the result. The alcoholmeter reading (*A*) is expressed in green in the horizontal column, the temperature (*T*) of the spirit is expressed in blue in the vertical column. For example, if (*A*) = 10% vol and (*T*) = 15 °C, the result is 10.8% vol.

Table 4.5 Tables of correction of alcoholic strength one-tenth of a degree at a temperature of 20 °C

(A)	0.0	0.1	0.2	0.3	0.4	0.5	0.6	0.7	0.8	0.9	1.0	1.1	1.2	1.3	1.4	1.5	1.6	1.7	1.8	1.9
(T)																				
6.0	0.9	1.0	1.1	1.2	1.3	1.4	1.5	1.6	1.7	1.8	1.9	2.0	2.1	2.2	2.4	2.5	2.6	2.7	2.8	2.9
6.5	0.9	1.0	1.1	1.2	1.3	1.4	1.5	1.6	1.7	1.8	1.9	2.0	2.1	2.2	2.4	2.5	2.6	2.7	2.8	2.9
7.0	0.9	1.0	1.1	1.2	1.3	1.4	1.5	1.6	1.7	1.8	1.9	2.0	2.1	2.2	2.3	2.4	2.6	2.7	2.8	2.9
7.5	0.9	1.0	1.1	1.2	1.3	1.4	1.5	1.6	1.7	1.8	1.9	2.0	2.1	2.2	2.3	2.4	2.5	2.6	2.7	2.9
8.0	0.9	1.0	1.1	1.2	1.3	1.4	1.5	1.6	1.7	1.8	1.9	2.0	2.1	2.2	2.3	2.4	2.5	2.6	2.7	2.8
8.5	0.9	1.0	1.1	1.2	1.3	1.4	1.5	1.6	1.7	1.8	1.9	2.0	2.1	2.2	2.3	2.4	2.5	2.6	2.7	2.8
9.0	0.9	1.0	1.1	1.2	1.3	1.4	1.5	1.6	1.7	1.8	1.9	2.0	2.1	2.2	2.3	2.4	2.5	2.6	2.7	2.8
9.5	0.8	0.9	1.0	1.1	1.2	1.3	1.5	1.6	1.7	1.8	1.9	2.0	2.1	2.2	2.3	2.4	2.5	2.6	2.7	2.8
10.0	0.8	0.9	1.0	1.1	1.2	1.3	1.4	1.5	1.6	1.7	1.8	1.9	2.0	2.2	2.3	2.4	2.5	2.6	2.7	2.8
10.5	0.8	0.9	1.0	1.1	1.2	1.3	1.4	1.5	1.6	1.7	1.8	1.9	2.0	2.1	2.2	2.3	2.4	2.5	2.6	2.7
11.0	0.8	0.9	1.0	1.1	1.2	1.3	1.4	1.5	1.6	1.7	1.8	1.9	2.0	2.1	2.2	2.3	2.4	2.5	2.6	2.7
11.5	0.7	0.8	1.0	1.1	1.2	1.3	1.4	1.5	1.6	1.7	1.8	1.9	2.0	2.1	2.2	2.3	2.4	2.5	2.6	2.7
12.0	0.7	0.8	0.9	1.0	1.1	1.2	1.3	1.4	1.5	1.6	1.7	1.8	1.9	2.0	2.1	2.3	2.4	2.5	2.6	2.7
12.5	0.7	0.8	0.9	1.0	1.1	1.2	1.3	1.4	1.5	1.6	1.7	1.8	1.9	2.0	2.1	2.2	2.3	2.4	2.5	2.6
13.0	0.7	0.8	0.9	1.0	1.1	1.2	1.3	1.4	1.5	1.6	1.7	1.8	1.9	2.0	2.1	2.2	2.3	2.4	2.5	2.6
13.5	0.6	0.7	0.8	0.9	1.0	1.1	1.2	1.3	1.4	1.5	1.6	1.7	1.8	1.9	2.0	2.1	2.3	2.4	2.5	2.6
14.0	0.6	0.7	0.8	0.9	1.0	1.1	1.2	1.3	1.4	1.5	1.6	1.7	1.8	1.9	2.0	2.1	2.2	2.3	2.4	2.5
14.5	0.5	0.6	0.7	0.9	1.0	1.1	1.2	1.3	1.4	1.5	1.6	1.7	1.8	1.9	2.0	2.1	2.2	2.3	2.4	2.5
15.0	0.5	0.6	0.7	0.8	0.9	1.0	1.1	1.2	1.3	1.4	1.5	1.6	1.7	1.8	1.9	2.0	2.1	2.2	2.3	2.4
15.5	0.5	0.6	0.7	0.8	0.9	1.0	1.1	1.2	1.3	1.4	1.5	1.6	1.7	1.8	1.9	2.0	2.1	2.2	2.3	2.4
16.0	0.4	0.5	0.6	0.7	0.8	0.9	1.0	1.1	1.2	1.3	1.4	1.5	1.6	1.7	1.8	1.9	2.0	2.1	2.2	2.3
16.5	0.4	0.5	0.6	0.7	0.8	0.9	1.0	1.1	1.2	1.3	1.4	1.5	1.6	1.7	1.8	1.9	2.0	2.1	2.2	2.3
17.0	0.3	0.4	0.5	0.6	0.7	0.8	0.9	1.0	1.1	1.2	1.3	1.4	1.5	1.6	1.7	1.8	1.9	2.0	2.1	2.2
17.5	0.3	0.4	0.5	0.6	0.7	0.8	0.9	1.0	1.1	1.2	1.3	1.4	1.5	1.6	1.7	1.8	1.9	2.0	2.1	2.2
18.0	0.2	0.3	0.4	0.5	0.6	0.7	0.8	0.9	1.0	1.1	1.2	1.3	1.4	1.5	1.6	1.7	1.8	1.9	2.0	2.1
18.5	0.2	0.3	0.4	0.5	0.6	0.7	0.8	0.9	1.0	1.1	1.2	1.3	1.4	1.5	1.6	1.7	1.8	1.9	2.0	2.1

(*A*)	0.0	0.1	0.2	0.3	0.4	0.5	0.6	0.7	0.8	0.9	1.0	1.1	1.2	1.3	1.4	1.5	1.6	1.7	1.8	1.9
(*T*)																				
19.0	0.1	0.2	0.3	0.4	0.5	0.6	0.7	0.8	0.9	1.0	1.1	1.2	1.3	1.4	1.5	1.6	1.7	1.8	1.9	2.0
19.5	0.1	0.2	0.3	0.4	0.5	0.6	0.7	0.8	0.9	1.0	1.1	1.2	1.3	1.4	1.5	1.6	1.7	1.8	1.9	2.0
20.0	0.0	0.1	0.2	0.3	0.4	0.5	0.6	0.7	0.8	0.9	1.0	1.1	1.2	1.3	1.4	1.5	1.6	1.7	1.8	1.9
20.5	–	0.0	0.1	0.2	0.3	0.4	0.5	0.6	0.7	0.8	0.9	1.0	1.1	1.2	1.3	1.4	1.5	1.6	1.7	1.8
21.0	–	–	0.1	0.2	0.3	0.4	0.5	0.6	0.7	0.8	0.9	1.0	1.1	1.2	1.3	1.4	1.5	1.6	1.7	1.8
21.5	–	–	0.0	0.1	0.2	0.3	0.4	0.5	0.6	0.7	0.8	0.9	1.0	1.1	1.2	1.3	1.4	1.5	1.6	1.7
22.0	–	–	–	0.0	0.1	0.2	0.3	0.4	0.5	0.6	0.7	0.8	0.9	1.0	1.1	1.2	1.3	1.4	1.5	1.6
22.5	–	–	–	–	0.1	0.2	0.3	0.4	0.5	0.6	0.7	0.8	0.9	1.0	1.1	1.2	1.3	1.4	1.5	1.6
23.0	–	–	–	–	0.0	0.1	0.2	0.3	0.4	0.5	0.6	0.7	0.8	0.9	1.0	1.1	1.2	1.3	1.4	1.5
23.5	–	–	–	–	–	0.0	0.1	0.2	0.3	0.4	0.5	0.6	0.7	0.8	0.9	1.0	1.1	1.2	1.3	1.4
24.0	–	–	–	–	–	–	0.1	0.2	0.3	0.4	0.5	0.6	0.7	0.8	0.9	1.0	1.1	1.2	1.2	1.3
24.5	–	–	–	–	–	–	0.0	0.1	0.2	0.3	0.4	0.5	0.6	0.7	0.8	0.9	1.0	1.1	1.2	1.3
25.0	–	–	–	–	–	–	–	0.0	0.1	0.2	0.3	0.4	0.5	0.6	0.7	0.8	0.9	1.0	1.1	1.2
25.5	–	–	–	–	–	–	–	–	0.0	0.1	0.2	0.3	0.4	0.5	0.6	0.7	0.8	0.9	1.0	1.1
26.0	–	–	–	–	–	–	–	–	–	0.1	0.2	0.3	0.4	0.4	0.5	0.6	0.7	0.8	0.9	1.0
26.5	–	–	–	–	–	–	–	–	–	0.0	0.1	0.2	0.3	0.4	0.5	0.6	0.7	0.8	0.9	1.0
27.0	–	–	–	–	–	–	–	–	–	–	0.0	0.1	0.2	0.3	0.4	0.5	0.6	0.7	0.8	0.9
27.5	–	–	–	–	–	–	–	–	–	–	–	0.0	0.1	0.2	0.3	0.4	0.5	0.6	0.7	0.8
28.0	–	–	–	–	–	–	–	–	–	–	–	–	0.0	0.1	0.2	0.3	0.4	0.5	0.6	0.7
28.5	–	–	–	–	–	–	–	–	–	–	–	–	–	0.0	0.1	0.2	0.3	0.4	0.5	0.6
29.0	–	–	–	–	–	–	–	–	–	–	–	–	–	–	0.0	0.1	0.2	0.3	0.4	0.5
29.5	–	–	–	–	–	–	–	–	–	–	–	–	–	–	–	0.0	0.1	0.2	0.3	0.4
30.0	–	–	–	–	–	–	–	–	–	–	–	–	–	–	–	–	0.1	0.1	0.2	0.3

(continued)

Table 4.5 (continued)

(*A*)	2.0	2.1	2.2	2.3	2.4	2.5	2.6	2.7	2.8	2.9	3.0	3.1	3.2	3.3	3.4	3.5	3.6	3.7	3.8	3.9
(*T*)																				
6.0	3.0	3.1	3.2	3.3	3.4	3.5	3.6	3.7	3.8	3.9	4.0	4.1	4.2	4.3	4.4	4.5	4.6	4.7	4.8	4.9
6.5	3.0	3.1	3.2	3.3	3.4	3.5	3.6	3.7	3.8	3.9	4.0	4.1	4.2	4.3	4.4	4.6	4.7	4.8	4.9	5.0
7.0	3.0	3.1	3.2	3.3	3.4	3.5	3.6	3.7	3.8	3.9	4.0	4.1	4.2	4.3	4.4	4.5	4.6	4.8	4.9	5.0
7.5	3.0	3.1	3.2	3.3	3.4	3.5	3.6	3.7	3.8	3.9	4.0	4.1	4.2	4.3	4.4	4.5	4.6	4.7	4.8	5.0
8.0	2.9	3.0	3.2	3.3	3.4	3.5	3.6	3.7	3.8	3.9	4.0	4.1	4.2	4.3	4.4	4.5	4.6	4.7	4.8	4.9
8.5	2.9	3.0	3.1	3.2	3.3	3.5	3.6	3.7	3.8	3.9	4.0	4.1	4.2	4.3	4.4	4.5	4.6	4.7	4.8	4.9
9.0	2.9	3.0	3.1	3.2	3.3	3.4	3.5	3.6	3.7	3.9	4.0	4.1	4.2	4.3	4.4	4.5	4.6	4.7	4.8	4.9
9.5	2.9	3.0	3.1	3.2	3.3	3.4	3.5	3.6	3.7	3.8	3.9	4.0	4.1	4.2	4.3	4.4	4.5	4.6	4.7	4.8
10.0	2.9	3.0	3.1	3.2	3.3	3.4	3.5	3.6	3.7	3.8	3.9	4.0	4.1	4.2	4.3	4.4	4.5	4.6	4.7	4.9
10.5	2.8	3.0	3.1	3.2	3.3	3.4	3.5	3.6	3.7	3.8	3.9	4.0	4.1	4.2	4.3	4.4	4.5	4.6	4.7	4.8
11.0	2.8	2.9	3.0	3.1	3.2	3.3	3.4	3.5	3.7	3.8	3.9	4.0	4.1	4.2	4.3	4.4	4.5	4.6	4.7	4.8
11.5	2.8	2.9	3.0	3.1	3.2	3.3	3.4	3.5	3.6	3.7	3.8	3.9	4.0	4.1	4.2	4.3	4.5	4.6	4.7	4.8
12.0	2.8	2.9	3.0	3.1	3.2	3.3	3.4	3.5	3.6	3.7	3.8	3.9	4.0	4.1	4.2	4.3	4.4	4.5	4.6	4.7
12.5	2.7	2.8	2.9	3.0	3.1	3.2	3.3	3.5	3.6	3.7	3.8	3.9	4.0	4.1	4.2	4.3	4.4	4.5	4.6	4.7
13.0	2.7	2.8	2.9	3.0	3.1	3.2	3.3	3.4	3.5	3.6	3.7	3.8	3.9	4.0	4.1	4.2	4.3	4.4	4.6	4.7
13.5	2.7	2.8	2.9	3.0	3.1	3.2	3.3	3.4	3.5	3.6	3.7	3.8	3.9	4.0	4.1	4.2	4.3	4.4	4.5	4.6
14.0	2.6	2.7	2.8	2.9	3.0	3.1	3.2	3.3	3.4	3.5	3.6	3.7	3.9	4.0	4.1	4.2	4.3	4.4	4.5	4.6
14.5	2.6	2.7	2.8	2.9	3.0	3.1	3.2	3.3	3.4	3.5	3.6	3.7	3.8	3.9	4.0	4.1	4.2	4.3	4.4	4.5
15.0	2.5	2.6	2.7	2.8	2.9	3.0	3.1	3.2	3.4	3.5	3.6	3.7	3.8	3.9	4.0	4.1	4.2	4.3	4.4	4.5
15.5	2.5	2.6	2.7	2.8	2.9	3.0	3.1	3.2	3.3	3.4	3.5	3.6	3.7	3.8	3.9	4.0	4.1	4.2	4.3	4.4
16.0	2.4	2.5	2.6	2.7	2.9	3.0	3.1	3.2	3.3	3.4	3.5	3.6	3.7	3.8	3.9	4.0	4.1	4.2	4.3	4.4
16.5	2.4	2.5	2.6	2.7	2.8	2.9	3.0	3.1	3.2	3.3	3.4	3.5	3.6	3.7	3.8	3.9	4.0	4.1	4.2	4.3
17.0	2.3	2.4	2.5	2.6	2.7	2.9	3.0	3.1	3.2	3.3	3.4	3.5	3.6	3.7	3.8	3.9	4.0	4.1	4.2	4.3
17.5	2.3	2.4	2.5	2.6	2.7	2.8	2.9	3.0	3.1	3.2	3.3	3.4	3.5	3.6	3.7	3.8	3.9	4.0	4.1	4.2
18.0	2.2	2.3	2.4	2.5	2.6	2.7	2.8	2.9	3.0	3.1	3.2	3.3	3.4	3.5	3.6	3.8	3.9	4.0	4.1	4.2
18.5	2.2	2.3	2.4	2.5	2.6	2.7	2.8	2.9	3.0	3.1	3.2	3.3	3.4	3.5	3.6	3.7	3.8	3.9	4.0	4.1

(A)	2.0	2.1	2.2	2.3	2.4	2.5	2.6	2.7	2.8	2.9	3.0	3.1	3.2	3.3	3.4	3.5	3.6	3.7	3.8	3.9
(T)																				
19.0	2.1	2.2	2.3	2.4	2.5	2.6	2.7	2.8	2.9	3.0	3.1	3.2	3.3	3.4	3.5	3.6	3.7	3.8	3.9	4.0
19.5	2.1	2.2	2.3	2.4	2.5	2.6	2.7	2.8	2.9	3.0	3.1	3.2	3.3	3.4	3.5	3.6	3.7	3.8	3.9	4.0
20.0	2.0	2.1	2.2	2.3	2.4	2.5	2.6	2.7	2.8	2.9	3.0	3.1	3.2	3.3	3.4	3.5	3.6	3.7	3.8	3.9
20.5	1.9	2.0	2.1	2.2	2.3	2.4	2.5	2.6	2.7	2.8	2.9	3.0	3.1	3.2	3.3	3.4	3.5	3.6	3.7	3.8
21.0	1.9	2.0	2.1	2.2	2.3	2.4	2.5	2.6	2.7	2.8	2.9	3.0	3.1	3.2	3.3	3.4	3.5	3.6	3.7	3.8
21.5	1.8	1.9	2.0	2.1	2.2	2.3	2.4	2.5	2.6	2.7	2.8	2.9	3.0	3.1	3.2	3.3	3.4	3.5	3.6	3.7
22.0	1.7	1.8	1.9	2.0	2.1	2.2	2.3	2.4	2.5	2.6	2.7	2.8	2.9	3.0	3.1	3.2	3.3	3.4	3.5	3.6
22.5	1.7	1.8	1.9	2.0	2.1	2.2	2.3	2.4	2.5	2.6	2.7	2.8	2.9	3.0	3.1	3.1	3.2	3.3	3.4	3.5
23.0	1.6	1.7	1.8	1.9	2.0	2.1	2.2	2.3	2.4	2.5	2.6	2.7	2.8	2.9	3.0	3.1	3.2	3.3	3.4	3.5
23.5	1.5	1.6	1.7	1.8	1.9	2.0	2.1	2.2	2.3	2.4	2.5	2.6	2.7	2.8	2.9	3.0	3.1	3.2	3.3	3.4
24.0	1.4	1.5	1.6	1.7	1.8	1.9	2.0	2.1	2.2	2.3	2.4	2.5	2.6	2.7	2.8	2.9	3.0	3.1	3.2	3.3
24.5	1.4	1.5	1.6	1.7	1.8	1.9	2.0	2.1	2.2	2.3	2.3	2.4	2.5	2.6	2.7	2.8	2.9	3.0	3.1	3.2
25.0	1.3	1.4	1.5	1.6	1.7	1.8	1.9	2.0	2.1	2.2	2.3	2.4	2.5	2.6	2.7	2.8	2.9	3.0	3.1	3.1
25.5	1.2	1.3	1.4	1.5	1.6	1.7	1.8	1.9	2.0	2.1	2.2	2.3	2.4	2.5	2.6	2.7	2.8	2.9	3.0	3.1
26.0	1.1	1.2	1.3	1.4	1.5	1.6	1.7	1.8	1.9	2.0	2.1	2.2	2.3	2.4	2.5	2.6	2.7	2.8	2.9	3.0
26.5	1.0	1.1	1.2	1.3	1.4	1.5	1.6	1.7	1.8	1.9	2.0	2.1	2.2	2.3	2.4	2.5	2.6	2.7	2.8	2.9
27.0	1.0	1.1	1.2	1.3	1.4	1.5	1.5	1.6	1.7	1.8	1.9	2.0	2.1	2.2	2.3	2.4	2.5	2.6	2.7	2.8
27.5	0.9	1.0	1.1	1.2	1.3	1.4	1.5	1.6	1.7	1.8	1.8	1.9	2.0	2.1	2.2	2.3	2.4	2.5	2.6	2.7
28.0	0.8	0.9	1.0	1.1	1.2	1.3	1.4	1.5	1.6	1.7	1.8	1.9	2.0	2.0	2.1	2.2	2.3	2.4	2.5	2.6
28.5	0.7	0.8	0.9	1.0	1.1	1.2	1.3	1.4	1.5	1.6	1.7	1.8	1.9	2.0	2.1	2.1	2.2	2.3	2.4	2.5
29.0	0.6	0.7	0.8	0.9	1.0	1.1	1.2	1.3	1.4	1.5	1.6	1.7	1.8	1.9	2.0	2.1	2.2	2.2	2.3	2.4
29.5	0.5	0.6	0.7	0.8	0.9	1.0	1.1	1.2	1.3	1.4	1.5	1.6	1.7	1.8	1.9	2.0	2.1	2.2	2.2	2.3
30.0	0.4	0.5	0.6	0.7	0.8	0.9	1.0	1.1	1.2	1.3	1.4	1.5	1.6	1.7	1.8	1.9	2.0	2.1	2.2	2.2

(continued)

Table 4.5 (continued)

(*A*)	4.0	4.1	4.2	4.3	4.4	4.5	4.6	4.7	4.8	4.9	5.0	5.1	5.2	5.3	5.4	5.5	5.6	5.7	5.8	5.9
(*T*)																				
6.0	5.1	5.2	5.3	5.4	5.5	5.6	5.7	5.8	5.9	6.1	6.2	6.3	6.4	6.5	6.6	6.7	6.8	6.9	7.0	7.1
6.5	5.1	5.2	5.3	5.4	5.5	5.6	5.7	5.8	5.9	6.0	6.2	6.3	6.4	6.5	6.6	6.7	6.8	6.9	7.0	7.1
7.0	5.1	5.2	5.3	5.4	5.5	5.6	5.7	5.8	5.9	6.0	6.1	6.3	6.4	6.5	6.6	6.7	6.8	6.9	7.0	7.1
7.5	5.1	5.2	5.3	5.4	5.5	5.6	5.7	5.8	5.9	6.0	6.1	6.2	6.3	6.5	6.6	6.7	6.8	6.9	7.0	7.1
8.0	5.0	5.2	5.3	5.4	5.5	5.6	5.7	5.8	5.9	6.0	6.1	6.2	6.3	6.4	6.5	6.7	6.8	6.9	7.0	7.1
8.5	5.0	5.1	5.2	5.3	5.5	5.6	5.7	5.8	5.9	6.0	6.1	6.2	6.3	6.4	6.5	6.6	6.7	6.8	7.0	7.1
9.0	5.0	5.1	5.2	5.3	5.4	5.5	5.6	5.7	5.9	6.0	6.1	6.2	6.3	6.4	6.5	6.6	6.7	6.8	6.9	7.0
9.5	5.0	5.1	5.2	5.3	5.4	5.5	5.6	5.7	5.8	5.9	6.0	6.1	6.3	6.4	6.5	6.6	6.7	6.8	6.9	7.0
10.0	5.0	5.1	5.2	5.3	5.4	5.5	5.6	5.7	5.8	5.9	6.0	6.1	6.2	6.3	6.4	6.5	6.7	6.8	6.9	7.0
10.5	4.9	5.0	5.1	5.2	5.4	5.5	5.6	5.7	5.8	5.9	6.0	6.1	6.2	6.3	6.4	6.5	6.6	6.7	6.8	6.9
11.0	4.9	5.0	5.1	5.2	5.3	5.4	5.5	5.6	5.7	5.8	6.0	6.1	6.2	6.3	6.4	6.5	6.6	6.7	6.8	6.9
11.5	4.9	5.0	5.1	5.2	5.3	5.4	5.5	5.6	5.7	5.8	5.9	6.0	6.1	6.2	6.3	6.4	6.6	6.7	6.8	6.9
12.0	4.8	4.9	5.0	5.1	5.3	5.4	5.5	5.6	5.7	5.8	5.9	6.0	6.1	6.2	6.3	6.4	6.5	6.6	6.7	6.8
12.5	4.8	4.9	5.0	5.1	5.2	5.3	5.4	5.5	5.6	5.7	5.8	5.9	6.1	6.2	6.3	6.4	6.5	6.6	6.7	6.8
13.0	4.8	4.9	5.0	51.0	5.2	5.3	5.4	5.5	5.6	5.7	5.8	5.9	6.0	6.1	6.2	6.3	6.4	6.5	6.6	6.7
13.5	4.7	4.8	4.9	5.0	5.1	5.2	5.3	5.4	5.5	5.7	5.8	5.9	6.0	6.1	6.2	6.3	6.4	6.5	6.6	6.7
14.0	4.7	4.8	4.9	5.0	5.1	5.2	5.3	5.4	5.5	5.6	5.7	5.8	5.9	6.0	6.1	6.2	6.3	6.4	6.5	6.6
14.5	4.6	4.7	4.8	4.9	5.0	5.1	5.2	5.4	5.5	5.6	5.7	5.8	5.9	6.0	6.1	6.2	6.3	6.4	6.5	6.6
15.0	4.6	4.7	4.8	4.9	5.0	5.1	5.2	5.3	5.4	5.5	5.6	5.7	5.8	5.9	6.0	6.1	6.2	6.3	6.4	6.5
15.5	4.5	4.6	4.7	4.8	4.9	5.0	5.1	5.2	5.4	5.5	5.6	5.7	5.8	5.9	6.0	6.1	6.2	6.3	6.4	6.5
16.0	4.5	4.6	4.7	4.8	4.9	5.0	5.1	5.2	5.3	5.4	5.5	5.6	5.7	5.8	5.9	6.0	6.1	6.2	6.3	6.4
16.5	4.4	4.5	4.6	4.7	4.8	4.9	5.0	5.1	5.2	5.3	5.4	5.5	5.7	5.8	5.9	6.0	6.1	6.2	6.3	6.4
17.0	4.4	4.5	4.6	4.7	4.8	4.9	5.0	5.1	5.2	5.3	5.4	5.5	5.6	5.7	5.8	5.9	6.0	6.1	6.2	6.3
17.5	4.3	4.4	4.5	4.6	4.7	4.8	4.9	5.0	5.1	5.2	5.3	5.4	5.5	5.6	5.7	5.8	5.9	6.0	6.1	6.2
18.0	4.3	4.4	4.5	4.6	4.7	4.8	4.9	5.0	5.1	5.2	5.3	5.4	5.5	5.6	5.7	5.8	5.9	6.0	6.1	6.2
18.5	4.2	4.3	4.4	4.5	4.6	4.7	4.8	4.9	5.0	5.1	5.2	5.3	5.4	5.5	5.6	5.7	5.8	5.9	6.0	6.1

(*A*)	4.0	4.1	4.2	4.3	4.4	4.5	4.6	4.7	4.8	4.9	5.0	5.1	5.2	5.3	5.4	5.5	5.6	5.7	5.8	5.9
(*T*)																				
19.0	4.1	4.2	4.3	4.4	4.5	4.6	4.7	4.8	4.9	5.0	5.1	5.2	5.3	5.4	5.5	5.6	5.7	5.8	5.9	6.0
19.5	4.1	4.2	4.3	4.4	4.5	4.6	4.7	4.8	4.9	5.0	5.1	5.2	5.3	5.4	5.5	5.6	5.7	5.8	5.9	6.0
20.0	4.0	4.1	4.2	4.3	4.4	4.5	4.6	4.7	4.8	4.9	5.0	5.1	5.2	5.3	5.4	5.5	5.6	5.7	5.8	5.9
20.5	3.9	4.0	4.1	4.2	4.3	4.4	4.5	4.6	4.7	4.8	4.9	5.0	5.1	5.2	5.3	5.4	5.5	5.6	5.7	5.8
21.0	3.9	4.0	4.1	4.2	4.3	4.4	4.5	4.6	4.7	4.8	4.9	5.0	5.1	5.2	5.3	5.4	5.5	5.6	5.7	5.8
21.5	3.8	3.9	4.0	4.1	4.2	4.3	4.4	4.5	4.6	4.7	4.8	4.9	5.0	5.1	5.2	5.3	5.4	5.5	5.6	5.7
22.0	3.7	3.8	3.9	4.0	4.1	4.2	4.3	4.4	4.5	4.6	4.7	4.8	4.9	5.0	5.1	5.2	5.3	5.4	5.5	5.6
22.5	3.6	3.7	3.8	3.9	4.0	4.1	4.2	4.3	4.4	4.5	4.6	4.7	4.8	4.9	5.0	5.1	5.2	5.3	5.4	5.5
23.0	3.6	3.7	3.8	3.9	4.0	4.1	4.2	4.3	4.4	4.5	4.5	4.6	4.7	4.8	4.9	5.0	5.1	5.2	5.3	5.4
23.5	3.5	3.6	3.7	3.8	3.9	4.0	4.1	4.2	4.3	4.4	4.5	4.6	4.7	4.8	4.9	5.0	5.1	5.2	5.3	5.3
24.0	3.4	3.5	3.6	3.7	3.8	3.9	4.0	4.1	4.2	4.3	4.4	4.5	4.6	4.7	4.8	4.9	5.0	5.1	5.2	5.3
24.5	3.3	3.4	3.5	3.6	3.7	3.8	3.9	4.0	4.1	4.2	4.3	4.4	4.5	4.6	4.7	4.8	4.9	5.0	5.1	5.2
25.0	3.2	3.3	3.4	3.5	3.6	3.7	3.8	3.9	4.0	4.1	4.2	4.3	4.4	4.5	4.6	4.7	4.8	4.9	5.0	5.1
25.5	3.2	3.3	3.4	3.5	3.5	3.6	3.7	3.8	3.9	4.0	4.1	4.2	4.3	4.4	4.5	4.6	4.7	4.8	4.9	5.0
26.0	3.1	3.2	3.3	3.4	3.5	3.6	3.7	3.8	3.8	3.9	4.0	4.1	4.2	4.3	4.4	4.5	4.6	4.7	4.8	4.9
26.5	3.0	3.1	3.2	3.3	3.4	3.5	3.6	3.7	3.8	3.9	4.0	4.0	4.1	4.2	4.3	4.4	4.5	4.6	4.7	4.8
27.0	2.9	3.0	3.1	3.2	3.3	3.4	3.5	3.6	3.7	3.8	3.9	4.0	4.1	4.1	4.2	4.3	4.4	4.5	4.6	4.7
27.5	2.8	2.9	3.0	3.1	3.2	3.3	3.4	3.5	3.6	3.7	3.8	3.9	4.0	4.1	4.2	4.2	4.3	4.4	4.5	4.6
28.0	2.7	2.8	2.9	3.0	3.1	3.2	3.3	3.4	3.5	3.6	3.7	3.8	3.9	4.0	4.1	4.2	4.2	4.3	4.4	4.5
28.5	2.6	2.7	2.8	2.9	3.0	3.1	3.2	3.3	3.4	3.5	3.6	3.7	3.8	3.9	4.0	4.1	4.1	4.2	4.3	4.4
29.0	2.5	2.6	2.7	2.8	2.9	3.0	3.1	3.2	3.3	3.4	3.5	3.6	3.7	3.8	3.9	4.0	4.1	4.1	4.2	4.3
29.5	2.4	2.5	2.6	2.7	2.8	2.9	3.0	3.1	3.2	3.3	3.4	3.5	3.6	3.7	3.8	3.9	4.0	4.0	4.1	4.2
30.0	2.3	2.4	2.5	2.6	2.7	2.8	2.9	3.0	3.1	3.2	3.3	3.4	3.5	3.6	3.7	3.8	3.9	3.9	4.0	4.1

(continued)

Table 4.5 (continued)

(A)	6.0	6.1	6.2	6.3	6.4	6.5	6.6	6.7	6.8	6.9	7.0	7.1	7.2	7.3	7.4	7.5	7.6	7.7	7.8	7.9
(T)																				
6.0	7.3	7.4	7.5	7.6	7.7	7.8	7.9	8.0	8.1	8.3	8.4	8.5	8.6	8.7	8.8	8.9	9.0	9.2	9.3	9.4
6.5	7.2	7.4	7.5	7.6	7.7	7.8	7.9	8.0	8.1	8.2	8.4	8.5	8.6	8.7	8.8	8.9	9.0	9.1	9.2	9.4
7.0	7.2	7.3	7.4	7.6	7.7	7.8	7.9	8.0	8.1	8.2	8.3	8.4	8.6	8.7	8.8	8.9	9.0	9.1	9.2	9.3
7.5	7.2	7.3	7.4	7.5	7.6	7.8	7.9	8.0	8.1	8.2	8.3	8.4	8.5	8.6	8.8	8.9	9.0	9.1	9.2	9.3
8.0	7.2	7.3	7.4	7.5	7.6	7.7	7.8	8.0	8.1	8.2	8.3	8.4	8.5	8.6	8.7	8.8	8.9	9.1	9.2	9.3
8.5	7.2	7.3	7.4	7.5	7.6	7.7	7.8	7.9	8.0	8.1	8.3	8.4	8.5	8.6	8.7	8.8	8.9	9.0	9.1	9.3
9.0	7.1	7.3	7.4	7.5	7.6	7.7	7.8	7.9	8.0	8.1	8.2	8.3	8.4	8.6	8.7	8.8	8.9	9.0	9.1	9.2
9.5	7.1	7.2	7.3	7.4	7.5	7.7	7.8	7.9	8.0	8.1	8.2	8.3	8.4	8.5	8.6	8.7	8.9	9.0	9.1	9.2
10.0	7.1	7.2	7.3	7.4	7.5	7.6	7.7	7.8	7.9	8.1	8.2	8.3	8.4	8.5	8.6	8.7	8.8	8.9	9.0	9.1
10.5	7.0	7.2	7.3	7.4	7.5	7.6	7.7	7.8	7.9	8.0	8.1	8.2	8.3	8.4	8.6	8.7	8.8	8.9	9.0	9.1
11.0	7.0	7.1	7.2	7.3	7.4	7.5	7.7	7.8	7.9	8.0	8.1	8.2	8.3	8.4	8.5	8.6	8.7	8.8	8.9	9.1
11.5	7.0	7.1	7.2	7.3	7.4	7.5	7.6	7.7	7.8	7.9	8.0	8.1	8.3	8.4	8.5	8.6	8.7	8.8	8.9	9.0
12.0	6.9	7.0	7.1	7.3	7.4	7.5	7.6	7.7	7.8	7.9	8.0	8.1	8.2	8.3	8.4	8.5	8.6	8.7	8.9	9.0
12.5	6.9	7.0	7.1	7.2	7.3	7.4	7.5	7.6	7.7	7.8	7.9	8.1	8.2	8.3	8.4	8.5	8.6	8.7	8.8	8.9
13.0	6.8	7.0	7.1	7.2	7.3	7.4	7.5	7.6	7.7	7.8	7.9	8.0	8.1	8.2	8.3	8.4	8.5	8.6	8.7	8.9
13.5	6.8	6.9	7.0	7.1	7.2	7.3	7.4	7.5	7.6	7.7	7.8	8.0	8.1	8.2	8.3	8.4	8.5	8.6	8.7	8.8
14.0	6.7	6.9	7.0	7.1	7.2	7.3	7.4	7.5	7.6	7.7	7.8	7.9	8.0	8.1	8.2	8.3	8.4	8.5	8.6	8.7
14.5	6.7	6.8	6.9	7.0	7.1	7.2	7.3	7.4	7.5	7.6	7.7	7.8	7.9	8.1	8.2	8.3	8.4	8.5	8.6	8.7
15.0	6.6	6.7	6.9	7.0	7.1	7.2	7.3	7.4	7.5	7.6	7.7	7.8	7.9	8.0	8.1	8.2	8.3	8.4	8.5	8.6
15.5	6.6	6.7	6.8	6.9	7.0	7.1	7.2	7.3	7.4	7.5	7.6	7.7	7.8	7.9	8.0	8.1	8.2	8.3	8.5	8.6
16.0	6.5	6.6	6.7	6.8	6.9	7.0	7.1	7.3	7.4	7.5	7.6	7.7	7.8	7.9	8.0	8.1	8.2	8.3	8.4	8.5
16.5	6.5	6.6	6.7	6.8	6.9	7.0	7.1	7.2	7.3	7.4	7.5	7.6	7.7	7.8	7.9	8.0	8.1	8.2	8.3	8.4
17.0	6.4	6.5	6.6	6.7	6.8	6.9	7.0	7.1	7.2	7.3	7.4	7.5	7.6	7.7	7.8	7.9	8.0	8.1	8.3	8.4
17.5	6.3	6.4	6.5	6.7	6.8	6.9	7.0	7.1	7.2	7.3	7.4	7.5	7.6	7.7	7.8	7.9	8.0	8.1	8.2	8.3
18.0	6.3	6.4	6.5	6.6	6.7	6.8	6.9	7.0	7.1	7.2	7.3	7.4	7.5	7.6	7.7	7.8	7.9	8.0	8.1	8.2
18.5	6.2	6.3	6.4	6.5	6.6	6.7	6.8	6.9	7.0	7.1	7.2	7.3	7.4	7.5	7.6	7.7	7.8	7.9	8.0	8.1

(*A*)	6.0	6.1	6.2	6.3	6.4	6.5	6.6	6.7	6.8	6.9	7.0	7.1	7.2	7.3	7.4	7.5	7.6	7.7	7.8	7.9
(*T*)																				
19.0	6.1	6.2	6.3	6.4	6.5	6.6	6.7	6.8	6.9	7.0	7.2	7.3	7.4	7.5	7.6	7.7	7.8	7.9	8.0	8.1
19.5	6.1	6.2	6.3	6.4	6.5	6.6	6.7	6.8	6.9	7.0	7.1	7.2	7.3	7.4	7.5	7.6	7.7	7.8	7.9	8.0
20.0	6.0	6.1	6.2	6.3	6.4	6.5	6.6	6.7	6.8	6.9	7.0	7.1	7.2	7.3	7.4	7.5	7.6	7.7	7.8	7.9
20.5	5.9	6.0	6.1	6.2	6.3	6.4	6.5	6.6	6.7	6.8	6.9	7.0	7.1	7.2	7.3	7.4	7.5	7.6	7.7	7.8
21.0	5.9	5.9	6.0	6.0.1	6.2	6.3	6.4	6.5	6.6	6.7	6.8	6.9	7.0	7.1	7.2	7.3	7.4	7.5	7.6	7.7
21.5	5.8	5.9	6.0	6.1	6.2	6.3	6.4	6.5	6.6	6.7	6.8	6.9	7.0	7.1	7.2	7.3	7.4	7.5	7.6	7.7
22.0	5.7	5.8	5.9	6.0	6.1	6.2	6.3	6.4	6.5	6.6	6.7	6.8	6.9	7.0	7.1	7.2	7.3	7.4	7.5	7.6
22.5	5.6	5.7	5.8	5.9	6.0	6.1	6.2	6.3	6.4	6.5	6.6	6.7	6.8	6.9	7.0	7.1	7.2	7.3	7.4	7.5
23.0	5.5	5.6	5.7	5.8	5.9	6.0	6.1	6.2	6.3	6.4	6.5	6.6	6.7	6.8	6.9	7.0	7.1	7.2	7.3	7.4
23.5	5.4	5.5	5.6	5.7	5.8	5.9	6.0	6.1	6.2	6.3	6.4	6.5	6.6	6.7	6.8	6.9	7.0	7.1	7.2	7.3
24.0	5.4	5.5	5.6	5.7	5.8	5.8	5.9	6.0	6.1	6.2	6.3	6.4	6.5	6.6	6.7	6.8	6.9	7.0	7.1	7.2
24.5	5.3	5.4	5.5	5.6	5.7	5.8	5.9	6.0	6.1	6.1	6.2	6.3	6.4	6.5	6.6	6.7	6.8	6.9	7.0	7.1
25.0	5.2	5.3	5.4	5.5	5.6	5.7	5.8	5.9	6.0	6.1	6.2	6.2	6.3	6.4	6.5	6.6	6.7	6.8	6.9	7.0
25.5	5.1	5.2	5.3	5.4	5.5	5.6	5.7	5.8	5.9	6.0	6.1	6.2	6.3	6.3	6.4	6.5	6.6	6.7	6.8	6.9
26.0	5.0	5.1	5.2	5.3	5.4	5.5	5.6	5.7	5.8	5.9	6.0	6.1	6.2	6.3	6.3	6.4	6.5	6.6	6.7	6.8
26.5	4.9	5.0	5.1	5.2	5.3	5.4	5.5	5.6	5.7	5.8	5.9	6.0	6.1	6.2	6.3	6.3	6.4	6.5	6.6	6.7
27.0	4.8	4.9	5.0	5.1	5.2	5.3	5.4	5.5	5.6	5.7	5.8	5.9	6.0	6.1	6.2	6.2	6.3	6.4	6.5	6.6
27.5	4.7	4.8	4.9	5.0	5.1	5.2	5.3	5.4	5.5	5.6	5.7	5.8	5.9	6.0	6.1	6.1	6.2	6.3	6.4	6.5
28.0	4.6	4.7	4.8	4.9	5.0	5.1	5.2	5.3	5.4	5.5	5.6	5.7	5.8	5.9	6.0	6.0	6.1	6.2	6.3	6.4
28.5	4.5	4.6	4.7	4.8	4.9	5.0	5.1	5.2	5.3	5.4	5.5	5.6	5.7	5.8	5.8	5.9	6.0	6.1	6.2	6.3
29.0	4.4	4.5	4.6	4.7	4.8	4.9	5.0	5.1	5.2	5.3	5.4	5.5	5.6	5.7	5.7	5.8	5.9	6.0	6.1	6.2
29.5	4.3	4.4	4.5	4.6	4.7	4.8	4.9	5.0	5.1	5.2	5.3	5.4	5.5	5.5	5.6	5.7	5.8	5.9	6.0	6.1
30.0	4.2	4.3	4.4	4.5	4.6	4.7	4.8	4.9	5.0	5.1	5.2	5.3	5.3	5.4	5.5	5.6	5.7	5.8	5.9	6.0

(continued)

Table 4.5 (continued)

(*A*)	8.0	8.1	8.2	8.3	8.4	8.5	8.6	8.7	8.8	8.9	9.0	9.1	9.2	9.3	9.4	9.5	9.6	9.7	9.8	9.9
(*T*)																				
6.0	9.5	9.6	9.7	9.8	9.9	10.1	10.2	10.3	10.4	10.5	10.6	10.8	10.9	11.0	11.1	11.2	11.3	11.5	11.6	11.7
6.5	9.5	9.6	9.7	9.8	9.9	10.0	10.2	10.3	10.4	10.5	10.6	10.7	10.8	11.0	11.1	11.2	11.3	11.4	11.5	11.7
7.0	9.4	9.6	9.7	9.8	9.9	10.0	10.1	10.2	10.4	10.5	10.6	10.7	10.8	10.9	11.0	11.2	11.3	11.4	11.5	11.6
7.5	9.4	9.5	9.6	9.8	9.9	10.0	10.1	10.2	10.3	10.4	10.6	10.7	10.8	10.9	11.0	11.1	11.2	11.4	11.5	11.6
8.0	9.4	9.5	9.6	9.7	9.8	10.0	10.1	10.2	10.3	10.4	10.5	10.6	10.7	10.9	11.0	11.1	11.2	11.3	11.4	11.5
8.5	9.4	9.5	9.6	9.7	9.8	9.9	10.0	10.1	10.3	10.4	10.5	10.6	10.7	10.8	10.9	11.0	11.2	11.3	11.4	11.5
9.0	9.3	9.4	9.5	9.7	9.8	9.9	10.0	10.1	10.2	10.3	10.4	10.6	10.7	10.8	10.9	11.0	11.1	11.2	11.3	11.5
9.5	9.3	9.4	9.5	9.6	9.7	9.8	10.0	10.1	10.2	10.3	10.4	10.5	10.6	10.7	10.8	11.0	11.1	11.2	11.3	11.4
10.0	9.2	9.4	9.5	9.6	9.7	9.8	9.9	10.0	10.1	10.2	10.4	10.5	10.6	10.7	10.8	10.9	11.0	11.1	11.2	11.4
10.5	9.2	9.3	9.4	9.5	9.6	9.8	9.9	10.0	10.1	10.2	10.3	10.4	10.5	10.6	10.7	10.9	11.0	11.1	11.2	11.3
11.0	9.2	9.3	9.4	9.5	9.6	9.7	9.8	9.9	10.0	10.1	10.3	10.4	10.5	10.6	10.7	10.8	10.9	11.0	11.1	11.2
11.5	9.1	9.2	9.3	9.4	9.5	9.7	9.8	9.9	10.0	10.1	10.2	10.3	10.4	10.5	10.6	10.7	10.9	11.0	11.1	11.2
12.0	9.1	9.2	9.3	9.4	9.5	9.6	9.7	9.8	9.9	10.0	10.1	10.3	10.4	10.5	10.6	10.7	10.8	10.9	11.0	11.1
12.5	9.0	9.1	9.2	9.3	9.4	9.6	9.7	9.8	9.9	10.0	10.1	10.2	10.3	10.4	10.5	10.6	10.7	10.8	11.0	11.1
13.0	9.0	9.1	9.2	9.3	9.4	9.5	9.6	9.7	9.8	9.9	10.0	10.1	10.2	10.4	10.5	10.6	10.7	10.8	10.9	11.0
13.5	8.9	9.0	9.1	9.2	9.3	9.4	9.5	9.6	9.8	9.9	10.0	10.1	10.2	10.3	10.4	10.5	10.6	10.7	10.8	10.9
14.0	8.8	9.0	9.1	9.2	9.3	9.4	9.5	9.6	9.7	9.8	9.9	10.0	10.1	10.2	10.3.	10.4	10.5	10.6	10.8	10.9
14.5	8.8	8.9	9.0	9.1	9.2	9.3	9.4	9.5	9.6	9.7	9.8	9.9	10.1	10.2	10.3	10.4	10.5	10.6	10.7	10.8
15.0	8.7	8.8	8.9	9.0	9.1	9.2	9.4	9.5	9.6	9.7	9.8	9.9	10.0	10.1	10.2	10.3	10.4	10.5	10.6	10.7
15.5	8.7	8.8	8.9	9.0	9.1	9.2	9.3	9.4	9.5	9.6	9.7	9.8	9.9	10.0	10.1	10.2	10.3	10.4	10.5	10.6
16.0	8.6	8.7	8.8	8.9	9.0	9.1	9.2	9.3	9.4	9.5	9.6	9.7	9.8	9.9	10.0	10.2	10.3	10.4	10.5	10.6
16.5	8.5	8.6	8.7	8.8	8.9	9.0	9.1	9.2	9.4	9.5	9.6	9.7	9.8	9.9	10.0	10.1	10.2	10.3	10.4	10.5
17.0	8.5	8.6	8.7	8.8	8.9	9.0	9.1	9.2	9.3	9.4	9.5	9.6	9.7	9.8	9.9	10.0	10.1	10.2	10.3	10.4
17.5	8.4	8.5	8.6	8.7	8.8	8.9	9.0	9.1	9.2	9.3	9.4	9.5	9.6	9.7	9.8	9.9	10.0	10.1	10.2	10.3
18.0	8.3	8.4	8.5	8.6	8.7	8.8	8.9	9.0	9.1	9.2	9.3	9.4	9.5	9.6	9.7	9.8	9.9	10.0	10.1	10.2
18.5	8.2	8.3	8.4	8.5	8.6	8.7	8.8	8.9	9.0	9.1	9.2	9.4	9.5	9.6	9.7	9.8	9.9	10.0	10.1	10.2

(*A*)	8.0	8.1	8.2	8.3	8.4	8.5	8.6	8.7	8.8	8.9	9.0	9.1	9.2	9.3	9.4	9.5	9.6	9.7	9.8	9.9
(*T*)																				
19.0	8.2	8.3	8.4	8.5	8.6	8.7	8.8	8.9	9.0	9.1	9.2	9.3	9.4	9.5	9.6	9.7	9.8	9.9	10.0	10.1
19.5	8.1	8.2	8.3	8.4	8.5	8.6	8.7	8.8	8.9	9.0	9.1	9.2	9.3	9.4	9.5	9.6	9.7	9.8	9.9	10.0
20.0	8.0	8.1	8.2	8.3	8.4	8.5	8.6	8.7	8.8	8.9	9.0	9.1	9.2	9.3	9.4	9.5	9.6	9.7	9.8	9.9
20.5	7.9	8.0	8.1	8.2	8.3	8.4	8.5	8.6	8.7	8.8	8.9	9.0	9.1	9.2	9.3	9.4	9.5	9.6	9.7	9.8
21.0	7.8	7.9	8.0	8.1	8.2	8.3	8.4	8.5	8.6	8.7	8.8	8.9	9.0	9.1	9.2	9.3	9.4	9.5	9.6	9.7
21.5	7.8	7.8	7.9	8.0	8.1	8.2	8.3	8.4	8.5	8.6	8.7	8.8	8.9	9.0	9.1	9.2	9.3	9.4	9.5	9.6
22.0	7.7	7.8	7.9	8.0	8.1	8.2	8.3	8.4	8.4	8.5	8.6	8.7	8.8	8.9	9.0	9.1	9.2	9.3	9.4	9.5
22.5	7.6	7.7	7.8	7.9	8.0	8.1	8.2	8.3	8.4	8.5	8.6	8.7	8.7	8.8	8.9	9.0	9.1	9.2	9.3	9.4
23.0	7.5	7.6	7.7	7.8	7.9	8.0	8.1	8.2	8.3	8.4	8.5	8.6	8.7	8.8	8.8	8.9	9.0	9.1	9.2	9.3
23.5	7.4	7.5	7.6	7.7	7.8	7.9	8.0	8.1	8.2	8.3	8.4	8.5	8.6	8.7	8.8	8.9	8.9	9.0	9.1	9.2
24.0	7.3	7.4	7.5	7.6	7.7	7.8	7.9	8.0	8.1	8.2	8.3	8.4	8.5	8.6	8.7	8.8	8.8	8.9	9.0	9.1
24.5	7.2	7.3	7.4	7.5	7.6	7.7	7.8	7.9	8.0	8.1	8.2	8.3	8.4	8.5	8.6	8.7	8.7	8.8	8.9	9.0
25.0	7.1	7.2	7.3	7.4	7.5	7.6	7.7	7.8	7.9	8.0	8.1	8.2	8.3	8.4	8.5	8.6	8.6	8.7	8.8	8.9
25.5	7.0	7.1	7.2	7.3	7.4	7.5	7.6	7.7	7.8	7.9	8.0	8.1	8.2	8.3	8.4	8.4	8.5	8.6	8.7	8.8
26.0	6.9	7.0	7.1	7.2	7.3	7.4	7.5	7.6	7.7	7.8	7.9	8.0	8.1	8.2	8.3	8.3	8.4	8.5	8.6	8.7
26.5	6.8	6.9	7.0	7.1	7.2	7.3	7.4	7.5	7.6	7.7	7.8	7.9	8.0	8.1	8.1	8.2	8.3	8.4	8.5	8.6
27.0	6.7	6.8	6.9	7.0	7.1	7.2	7.3	7.4	7.5	7.6	7.7	7.8	7.9	7.9	8.0	8.1	8.2	8.3	8.4	8.5
27.5	6.6	6.7	6.8	6.9	7.0	7.1	7.2	7.3	7.4	7.5	7.6	7.7	7.7	7.8	7.9	8.0	8.1	8.2	8.3	8.4
28.0	6.5	6.6	6.7	6.8	6.9	7.0	7.1	7.2	7.3	7.4	7.5	7.5	7.6	7.7	7.8	7.9	8.0	8.1	8.2	8.3
28.5	6.4	6.5	6.6	6.7	6.8	6.9	7.0	7.1	7.2	7.3	7.3	7.4	7.5	7.6	7.7	7.8	7.9	8.0	8.1	8.2
29.0	6.3	6.4	6.5	6.6	6.7	6.8	6.9	7.0	7.1	7.1	7.2	7.3	7.4	7.5	7.6	7.7	7.8	7.9	8.0	8.1
29.5	6.2	6.3	6.4	6.5	6.6	6.7	6.8	6.8	6.9	7.0	7.1	7.2	7.3	7.4	7.5	7.6	7.7	7.8	7.9	8.0
30.0	6.1	6.2	6.3	6.4	6.5	6.6	6.6	6.7	6.8	6.9	7.0	7.1	7.2	7.3	7.4	7.5	7.6	7.7	7.8	7.8

(continued)

Table 4.5 (continued)

(*A*)	10.0	10.1	10.2	10.3	10.4	10.5	10.6	10.7	10.8	10.9	11.0	11.1	11.2	11.3	11.4	11.5	11.6	11.7	11.8	11.9
(*T*)																				
6.0	11.8	11.9	12.1	12.2	12.3	12.4	12.5	12.6	12.8	12.9	13.0	13.1	13.3	13.4	13.5	13.6	13.7	13.9	14.0	14.1
6.5	11.8	11.9	12.0	12.1	12.3	12.4	12.5	12.6	12.7	12.8	13.0	13.1	13.2	13.3	13.4	13.6	13.7	13.8	13.9	14.1
7.0	11.7	11.9	12.0	12.1	12.2	12.3	12.4	12.6	12.7	12.8	12.9	13.0	13.2	13.3	13.4	13.5	13.6	13.8	13.9	14.0
7.5	11.7	11.8	11.9	12.1	12.2	12.3	12.4	12.5	12.6	12.8	12.9	13.0	13.1	13.2	13.3	13.5	13.6	13.7	13.8	13.9
8.0	11.7	11.8	11.9	12.0	12.1	12.2	12.4	12.5	12.6	12.7	12.8	12.9	13.1	13.2	13.3	13.4	13.5	13.6	13.8	13.9
8.5	11.6	11.7	11.8	12.0	12.1	12.2	12.3	12.4	12.5	12.7	12.8	12.9	13.0	13.1	13.2	13.4	13.5	13.6	13.7	13.8
9.0	11.6	11.7	11.8	11.9	12.0	12.1	12.3	12.4	12.5	12.6	12.7	12.8	12.9	13.1	13.2	13.3	13.4	13.5	13.6	13.8
9.5	11.5	11.6	11.7	11.9	12.0	12.1	12.2	12.3	12.4	12.5	12.7	12.8	12.9	13.0	13.1	13.2	13.3	13.5	13.6	13.7
10.0	11.5	11.6	11.7	11.8	11.9	12.0	12.1	12.3	12.4	12.5	12.6	12.7	12.8	12.9	13.1	13.2	13.3	13.4	13.5	13.6
10.5	11.4	11.5	11.6	11.7	11.9	12.0	12.1	12.2	12.3	12.4	12.5	12.6	12.8	12.9	13.0	13.1	13.2	13.3	13.4	13.6
11.0	11.4	11.5	11.6	11.7	11.8	11.9	12.0	12.1	12.2	12.4	12.5	12.6	12.7	12.8	12.9	13.0	13.1	13.3	13.4	13.5
11.5	11.3	11.4	11.5	11.6	11.7	11.8	12.0	12.1	12.2	12.3	12.4	12.5	12.6	12.7	12.9	13.0	13.1	13.2	13.3	13.4
12.0	11.2	11.3	11.5	11.6	11.7	11.8	11.9	12.0	12.1	12.2	12.3	12.4	12.6	12.7	12.8	12.9	13.0	13.1	13.2	13.3
12.5	11.2	11.3	11.4	11.5	11.6	11.7	11.8	11.9	12.0	12.2	12.3	12.4	12.5	12.6	12.7	12.8	12.9	13.0	13.1	13.3
13.0	11.1	11.2	11.3	11.4	11.5	11.6	11.8	11.9	12.0	12.1	12.2	12.3	12.4	12.5	12.6	12.7	12.8	13.0	13.1	13.2
13.5	11.0	11.1	11.3	11.4	11.5	11.6	11.7	11.8	11.9	12.0	12.1	12.2	12.3	12.4	12.6	12.7	12.8	12.9	13.0	13.1
14.0	11.0	11.1	11.2	11.3	11.4	11.5	11.6	11.7	11.8	11.9	12.0	12.1	12.3	12.4	12.5	12.6	12.7	12.8	12.9	13.0
14.5	10.9	11.0	11.1	11.2	11.3	11.4	11.5	11.6	11.8	11.9	12.0	12.1	12.2	12.3	12.4	12.5	12.6	12.7	12.8	12.9
15.0	10.8	10.9	11.0	11.1	11.2	11.4	11.5	11.6	11.7	11.8	11.9	12.0	12.1	12.2	12.3	12.4	12.5	12.6	12.7	12.8
15.5	10.8	10.9	11.0	11.1	11.2	11.3	11.4	11.5	11.6	11.7	11.8	11.9	12.0	12.1	12.2	12.3	12.4	12.5	12.6	12.8
16.0	10.7	10.8	10.9	11.0	11.1	11.2	11.3	11.4	11.5	11.6	11.7	11.8	11.9	12.0	12.1	12.2	12.4	12.5	12.6	12.7
16.5	10.6	10.7	10.8	10.9	11.0	11.1	11.2	11.3	11.4	11.5	11.6	11.7	11.8	11.9	12.1	12.2	12.3	12.4	12.5	12.6
17.0	10.5	10.6	10.7	10.8	10.9	11.0	11.1	11.2	11.3	11.4	11.5	11.7	11.8	11.9	12.0	12.1	12.2	12.3	12.4	12.5
17.5	10.4	10.5	10.6	10.7	10.8	10.9	11.1	11.2	11.3	11.4	11.5	11.6	11.7	11.8	11.9	12.0	12.1	1.2	12.3	12.4
18.0	10.3	10.5	10.6	10.7	10.8	10.9	11.0	11.1	11.2	11.3	11.4	11.5	11.6	11.7	11.8	11.9	12.0	12.1	12.2	12.3
18.5	10.3	10.4	10.5	10.6	10.7	10.8	10.9	11.0	11.1	11.2	11.3	11.4	11.5	11.6	11.7	11.8	11.9	12.0	12.1	12.2

(*A*)	10.0	10.1	10.2	10.3	10.4	10.5	10.6	10.7	10.8	10.9	11.0	11.1	11.2	11.3	11.4	11.5	11.6	11.7	11.8	11.9
(*T*)																				
19.0	10.2	10.3	10.4	10.5	10.6	10.7	10.8	10.9	11.0	11.1	11.2	11.3	11.4	11.5	11.6	11.7	11.8	11.9	12.0	12.1
19.5	10.1	10.2	10.3	10.4	10.5	10.6	10.7	10.8	10.9	11.0	11.1	11.2	11.3	11.4	11.5	11.6	11.7	11.8	11.9	12.0
20.0	10.0	10.1	10.2	10.3	10.4	10.5	10.6	10.7	10.8	10.9	11.0	11.1	11.2	11.3	11.4	11.5	11.6	11.7	11.8	11.9
20.5	9.9	10.0	10.1	10.2	10.3	10.4	10.5	10.6	10.7	10.8	10.9	11.0	11.1	11.2	11.3	11.4	11.5	11.6	11.7	11.8
21.0	9.8	9.9	10.0	10.1	10.2	10.3	10.4	10.5	10.6	10.7	10.8	10.9	11.0	11.1	11.2	11.3	11.4	11.5	11.6	11.7
21.5	9.7	9.8	9.9	10.0	10.1	10.2	10.3	10.4	10.5	10.6	10.7	10.8	10.9	11.0	11.1	11.2	11.3	11.4	11.5	11.6
22.0	9.6	9.7	9.8	9.9	10.0	10.1	10.2	10.3	10.4	10.5	10.6	10.7	10.8	10.9	11.0	11.1	11.2	11.3	11.4	11.5
22.5	9.5	9.6	9.7	9.8	9.9	10.0	10.1	10.2	10.3	10.4	10.5	10.6	10.7	10.8	10.9	11.0	11.1	11.2	11.3	11.4
23.0	9.4	9.5	9.6	9.7	9.8	9.9	10.0	10.1	10.2	10.3	10.4	10.5	10.6	10.7	10.8	10.9	11.0	11.1	11.2	11.3
23.5	9.3	9.4	9.5	9.6	9.7	9.8	9.9	10.0	10.1	10.2	10.3	10.4	10.5	10.6	10.7	10.8	10.9	11.0	11.1	11.2
24.0	9.2	9.3	9.4	9.5	9.6	9.7	9.8	9.9	10.0	10.1	10.2	10.3	10.4	10.5	10.6	10.7	10.8	10.9	11.0	11.1
24.5	9.1	9.2	9.3	9.4	9.5	9.6	9.7	9.8	9.9	10.0	10.1	10.2	10.3	10.4	10.5	10.6	10.7	10.8	10.8	10.9
25.0	9.0	9.1	9.2	9.3	9.4	9.5	9.6	9.7	9.8	9.9	10.0	10.1	10.2	10.3	10.4	10.5	10.5	10.6	10.7	10.8
25.5	8.9	9.0	9.1	9.2	9.3	9.4	9.5	9.6	9.7	9.8	9.9	10.0	10.1	10.2	10.2	10.3	10.4	10.5	10.6	10.7
26.0	8.8	8.9	9.0	9.1	9.2	9.3	9.4	9.5	9.6	9.7	9.8	9.9	10.0	10.0	10.1	10.2	10.3	10.4	10.5	10.6
26.5	8.7	8.8	8.9	9.0	9.1	9.2	9.3	9.4	9.5	9.6	9.7	9.7	9.8	9.9	10.0	10.1	10.2	10.3	10.4	10.5
27.0	8.6	8.7	8.8	8.9	9.0	9.1	9.2	9.3	9.4	9.4	9.5	9.6	9.7	9.8	9.9	10.0	10.1	10.2	10.3	10.4
27.5	8.5	8.6	8.7	8.8	8.9	9.0	9.1	9.1	9.2	9.3	9.4	9.5	9.6	9.7	9.8	9.9	10.0	10.1	10.2	10.3
28.0	8.4	8.5	9.6	8.7	8.8	8.8	8.9	9.0	9.1	9.2	9.3	9.4	9.5	9.6	9.7	9.8	9.9	10.0	10.1	10.1
28.5	8.3	8.4	8.5	8.6	8.6	8.7	8.8	8.9	9.0	9.1	9.2	9.3	9.4	9.5	9.6	9.7	9.7	9.8	9.9	10.0
29.0	8.2	8.3	8.3	8.4	8.5	8.6	8.7	8.8	8.9	9.0	9.1	9.2	9.3	9.4	9.4	9.5	9.6	9.7	9.8	9.9
29.5	8.0	8.1	8.2	8.3	8.4	8.5	8.6	8.7	8.8	8.9	9.0	9.1	9.1	9.2	9.3	9.4	9.5	9.6	9.7	9.8
30.0	7.9	8.0	8.1	8.2	8.3	8.4	8.5	8.6	8.7	8.8	8.8	8.9	9.0	9.1	9.2	9.3	9.4	9.5	9.6	9.7

(continued)

Table 4.5 (continued)

(*A*)	12.0	12.1	12.2	12.3	12.4	12.5	12.6	12.7	12.8	12.9	13.0	13.1	13.2	13.3	13.4	13.5	13.6	13.7	13.8	13.9
(*T*)																				
6.0	14.2	14.4	14.5	14.6	14.7	14.8	15.0	15.1	15.2	15.3	15.5	15.6	15.7	15.8	16.0	16.1	16.2	16.4	16.5	16.6
6.5	14.2	14.3	14.4	14.5	14.7	14.8	14.9	15.0	15.2	15.3	15.4	15.5	15.7	15.8	15.9	16.0	16.2	16.3	16.4	16.5
7.0	14.1	14.2	14.4	14.5	14.6	14.7	14.8	15.0	15.1	15.2	15.3	15.5	15.6	15.7	158.0	16.0	16.1	16.2	16.3	16.5
7.5	14.1	14.2	14.3	14.4	14.5	14.7	14.8	14.9	15.0	15.1	15.3	15.4	15.5	15.6	15.8	15.9	16.0	16.1	16.3	16.4
8.0	14.0	14.1	14.2	14.4	14.5	14.6	14.7	148.0	15.0	15.1	15.2	15.3	15.4	15.6	15.7	15.8	15.9	16.0	16.2	16.3
8.5	13.9	14.1	14.2	14.3	14.4	14.5	14.7	14.8	14.9	15.0	15.1	15.2	15.4	15.5	15.6	15.7	15.8	16.0	16.1	16.2
9.0	13.9	14.0	14.1	14.2	14.3	14.5	14.6	14.7	14.8	14.9	15.1	15.2	15.3	15.4	15.5	15.6	15.8	15.9	16.0	16.1
9.5	13.8	13.9	14.0	14.2	14.3	14.4	14.5	14.6	14.7	14.9	15.0	15.1	15.2	15.3	15.4	15.6	15.7	15.8	15.9	16.0
10.0	13.7	13.9	14.0	14.1	14.2	14.3	14.4	14.5	14.7	14.8	14.9	15.0	15.1	15.2	15.4	15.5	15.6	15.7	15.8	15.9
10.5	13.7	13.8	13.9	14.0	14.1	14.2	14.4	14.5	14.6	14.7	14.8	14.9	15.0	15.2	15.3	15.4	15.5	15.6	15.7	15.9
11.0	13.6	13.7	13.8	13.9	14.1	14.2	14.3	14.4	14.5	14.6	14.7	14.8	15.0	15.1	15.2	15.3	15.4	15.5	15.7	15.8
11.5	13.5	13.6	13.7	13.9	14.0	14.1	14.2	14.3	14.4	14.5	14.6	14.8	14.9	15.0	15.1	15.2	15.3	15.4	15.6	15.7
12.0	13.4	13.6	13.7	13.8	13.9	14.0	14.1	14.2	14.3	14.5	14.6	14.7	14.8	14.9	15.0	15.1	15.2	15.4	15.5	15.6
12.5	13.4	13.5	13.6	13.7	13.8	13.9	14.0	14.1	14.3	14.4	14.5	14.6	14.7	14.8	14.9	15.0	15.1	15.3	15.4	15.5
13.0	13.3	13.4	13.5	13.6	13.7	13.8	13.9	14.1	14.2	14.3	14.4	14.5	14.6	14.7	14.8	14.9	15.1	15.2	15.3	15.4
13.5	13.2	13.3	13.4	13.5	13.6	13.7	13.9	14.0	14.1	14.2	14.3	14.4	14.5	14.6	14.7	14.8	15.0	15.1	15.2	15.3
14.0	13.1	13.2	13.3	13.4	13.6	13.7	13.8	13.9	14.0	14.1	14.2	14.3	14.4	14.5	14.6	14.7	14.9	15.0	15.1	15.2
14.5	13.0	13.1	13.2	13.4	13.5	13.6	13.7	13.8	13.9	14.0	14.1	14.2	14.3	14.4	14.5	14.7	14.8	14.9	15.0	15.1
15.0	12.9	13.1	13.2	13.3	13.4	13.5	13.6	13.7	13.8	13.9	14.0	14.1	14.2	14.3	14.4	14.6	14.7	14.8	14.9	15.0
15.5	12.9	13.0	13.1	13.2	13.3	13.4	13.5	13.6	13.7	13.8	13.9	14.0	14.1	14.2	14.3	14.5	14.6	14.7	14.8	14.9
16.0	12.8	12.9	13.0	13.1	13.2	13.3	13.4	13.5	13.6	13.7	13.8	13.9	14.0	14.1	14.2	14.4	14.5	14.6	14.7	14.8
16.5	12.7	12.8	12.9	13.0	13.1	13.2	13.3	13.4	13.5	13.6	13.7	13.8	13.9	14.0	14.1	14.3	14.4	14.5	14.6	14.7
17.0	12.6	12.7	12.8	12.9	13.0	13.1	13.2	13.3	13.4	13.5	13.6	13.7	13.8	13.9	14.0	14.1	14.3	14.4	14.5	14.6
17.5	12.5	12.6	12.7	12.8	12.9	13.0	13.1	13.2	13.3	13.4	13.5	13.6	13.7	13.8	13.9	14.0	14.1	14.2	14.4	14.5
18.0	12.4	12.5	12.6	12.7	12.8	12.9	13.0	13.1	13.2	13.3	13.4	13.5	13.6	13.7	13.8	13.9	14.0	14.1	14.2	14.3
18.5	12.3	12.4	12.5	12.6	12.7	12.8	12.9	13.0	13.1	13.2	13.3	13.4	13.5	13.6	13.7	13.8	13.9	14.0	14.1	14.2

(*A*)	12.0	12.1	12.2	12.3	12.4	12.5	12.6	12.7	12.8	12.9	13.0	13.1	13.2	13.3	13.4	13.5	13.6	13.7	13.8	13.9
(*T*)																				
19.0	12.2	12.3	12.4	12.5	12.6	12.7	12.8	12.9	13.0	1.1	13.2	13.3	13.4	13.5	13.6	13.7	13.8	13.9	14.0	14.1
19.5	12.1	12.2	12.3	12.4	12.5	12.6	12.7	12.8	12.9	13.0	13.1	13.2	13.3	13.4	13.5	13.6	13.7	13.8	13.9	14.0
20.0	12.0	12.1	12.2	12.3	12.4	12.5	12.6	12.7	12.8	12.9	13.0	13.1	13.2	13.3	13.4	13.5	13.6	13.7	13.8	13.9
20.5	11.9	12.0	12.1	12.2	12.3	12.4	12.5	12.6	12.7	12.8	12.9	13.0	13.1	13.2	13.3	13.4	13.5	13.6	13.7	13.8
21.0	11.8	11.9	12.0	12.1	12.2	12.3	12.4	12.5	12.6	12.7	12.8	12.9	13.0	13.1	13.2	13.3	13.4	13.5	13.6	13.7
21.5	11.7	11.8	11.9	12.0	12.1	12.2	12.3	12.4	12.5	12.6	12.7	12.8	12.9	13.0	13.1	13.2	13.3	13.4	13.5	13.6
22.0	11.6	11.7	11.8	11.9	12.0	12.1	12.2	12.3	12.4	12.5	12.6	12.7	12.8	12.9	13.0	13.0	13.1	13.2	13.3	13.4
22.5	11.5	11.6	11.7	11.8	11.9	12.0	12.1	12.2	12.3	12.4	12.4	12.5	12.6	12.7	12.8	12.9	13.0	13.1	13.2	13.3
23.0	11.4	11.5	11.6	11.7	11.8	11.9	11.9	12.0	12.1	12.2	12.3	12.4	12.5	12.6	12.7	12.8	12.9	13.0	13.1	13.2
23.5	11.3	11.4	11.5	11.5	11.6	11.7	11.8	11.9	12.0	12.1	12.2	12.3	12.4	12.5	12.6	12.7	12.8	12.9	13.0	13.1
24.0	11.2	11.2	11.3	11.4	11.5	11.6	11.7	11.8	11.9	12.0	12.1	12.2	12.3	12.4	12.5	12.6	12.7	12.8	12.9	13.0
24.5	11.0	11.1	11.2	11.3	11.4	11.5	11.6	11.7	11.8	11.9	12.0	12.1	12.2	12.3	12.4	12.5	12.6	12.7	12.7	12.8
25.0	10.9	11.0	11.1	11.2	11.3	11.4	11.5	11.6	11.7	11.8	11.9	12.0	12.1	12.2	12.3	12.3	12.4	12.5	12.6	12.7
25.5	10.8	10.9	11.0	11.1	11.2	11.3	11.4	11.5	11.6	11.7	11.8	11.8	11.9	12.0	12.1	12.2	12.3	12.4	12.5	12.6
26.0	10.7	10.8	10.9	11.0	11.1	11.2	11.3	11.4	11.4	11.5	11.6	11.7	11.8	11.9	12.0	12.1	12.2	12.3	12.4	12.5
26.5	10.6	10.7	10.8	10.9	11.0	11.1	11.1	11.2	11.3	11.4	11.5	11.6	11.7	11.8	11.9	12.0	12.1	12.2	12.3	12.4
27.0	10.5	10.6	10.7	10.7	10.8	10.9	11.0	11.1	11.2	11.3	11.4	11.5	11.6	11.7	11.8	11.9	12.0	12.0	12.1	12.2
27.5	10.4	10.4	10.5	10.6	10.7	10.8	10.9	11.0	11.1	11.2	11.3	11.4	11.5	11.6	11.6	11.7	11.8	11.9	12.0	12.1
28.0	10.2	10.3	10.4	10.5	10.6	10.7	10.8	10.9	11.0	11.1	11.2	11.2	11.3	11.4	11.5	11.6	11.7	11.8	11.9	12.0
28.5	10.1	10.2	10.3	10.4	10.5	10.6	10.7	10.8	10.8	10.9	11.0	11.1	11.2	11.3	11.4	11.5	11.6	11.7	11.8	11.8
29.0	10.0	10.1	10.2	10.3	10.4	10.5	10.5	10.6	10.7	10.8	10.9	11.0	11.1	11.2	11.3	11.4	11.4	11.5	11.6	11.7
29.5	9.9	10.0	10.1	10.1	10.2	10.3	10.4	10.5	10.6	10.7	10.8	10.9	11.0	11.1	11.1	11.2	11.3	11.4	11.5	11.6
30.0	9.8	9.8	9.9	10.0	10.1	10.2	10.3	10.4	10.5	10.6	10.7	10.7	10.8	10.9	11.0	11.1	11.2	11.3	11.4	11.5

(continued)

Table 4.5 (continued)

(*A*)	14.0	14.1	14.2	14.3	14.4	14.5	14.6	14.7	14.8	14.9	15.0	15.1	15.2	15.3	15.4	15.5	15.6	15.7	15.8	15.9
(*T*)																				
6.0	16.7	16.9	17.0	17.1	17.2	17.4	17.5	17.6	17.8	17.9	18.0	18.1	18.3	18.4	18.5	18.7	18.8	18.9	19.1	19.2
6.5	16.7	16.8	16.9	17.0	17.2	17.3	17.4	17.5	17.7	17.8	17.9	18.1	18.2	18.3	18.4	18.6	18.7	18.8	19.0	19.1
7.0	16.6	16.7	16.8	17.0	17.1	17.2	17.3	17.5	17.6	17.7	17.8	18.0	18.1	18.2	18.3	18.5	18.6	18.7	18.8	19.0
7.5	16.5	16.6	16.7	16.9	17.0	17.1	17.2	17.4	17.5	17.6	17.7	17.9	18.0	18.1	18.2	18.4	18.5	18.6	18.7	18.9
8.0	16.4	16.5	16.7	16.8	16.9	17.0	17.1	17.3	17.4	17.5	17.6	17.8	17.9	18.0	18.1	18.3	18.4	18.5	18.6	18.8
8.5	16.3	16.4	16.6	16.7	16.8	16.9	17.1	17.2	17.3	17.4	17.5	17.7	17.8	17.9	18.0	18.2	18.3	18.4	18.5	18.6
9.0	16.2	16.4	16.5	16.6	16.7	16.8	17.0	17.1	17.2	17.3	17.4	17.6	17.7	17.8	17.9	18.0	18.2	18.3	18.4	18.5
9.5	16.2	16.3	16.4	16.5	16.6	16.7	16.9	17.0	17.1	17.2	17.3	17.5	17.6	17.7	17.8	17.9	18.1	18.2	18.3	18.4
10.0	16.1	16.2	16.3	16.4	16.5	16.7	16.8	16.9	17.0	17.1	17.2	17.4	17.5	17.6	17.7	17.8	18.0	18.1	18.2	18.3
10.5	16.0	16.1	16.2	16.3	16.4	16.6	16.7	16.8	16.9	17.0	17.1	17.3	17.4	17.5	17.6	17.7	17.8	18.0	18.1	18.2
11.0	15.9	16.0	16.1	16.2	16.3	16.5	16.6	16.7	16.8	16.9	17.0	17.2	17.3	17.4	17.5	17.6	17.7	17.8	18.0	18.1
11.5	15.8	15.9	16.0	16.1	16.2	16.4	16.5	16.6	16.7	16.8	16.9	17.0	17.2	17.3	17.4	17.5	17.6	17.7	17.8	18.0
12.0	15.7	15.8	15.9	16.0	16.1	16.3	16.4	16.5	16.6	16.7	16.8	16.9	17.1	17.2	17.3	17.4	17.5	17.6	17.7	17.8
12.5	15.6	15.7	15.8	15.9	16.0	16.2	16.3	16.4	16.5	16.6	16.7	16.8	16.9	17.1	17.2	17.3	17.4	17.5	17.6	17.7
13.0	15.5	15.6	15.7	15.8	15.9	16.1	16.2	16.3	16.4	16.5	16.6	16.7	16.8	16.9	17.1	17.2	17.3	17.4	17.5	17.6
13.5	15.4	15.5	15.6	15.7	15.8	15.9	16.1	16.2	16.3	16.4	16.5	16.6	16.7	16.8	16.9	17.1	17.2	17.3	17.4	17.5
14.0	15.3	15.4	15.5	15.6	15.7	15.8	16.0	16.1	16.2	16.3	16.4	16.5	16.6	16.7	16.8	16.9	17.0	17.2	17.3	17.4
14.5	15.2	15.3	15.4	15.5	15.6	15.7	15.8	16.0	16.1	16.2	16.3	16.4	16.5	16.6	16.7	16.8	16.9	17.0	17.1	17.3
15.0	15.1	15.2	15.3	15.4	15.5	15.6	15.7	15.8	16.0	16.1	16.2	163.0	16.4	16.5	16.6	16.7	16.8	16.9	17.0	17.1
15.5	15.0	15.1	15.2	15.3	15.4	15.5	15.6	15.7	15.8	15.9	16.1	16.2	16.3	16.4	16.5	16.6	16.7	16.8	16.9	17.0
16.0	14.9	15.0	15.1	15.2	15.3	15.4	15.5	15.6	15.7	15.8	15.9	16.0	16.2	16.3	16.4	16.5	16.6	16.7	16.8	16.9
16.5	14.8	14.9	15.0	15.1	15.2	15.3	15.4	15.5	15.6	15.7	15.8	15.9	16.0	16.1	16.2	16.4	16.5	16.6	16.7	16.8
17.0	14.7	14.8	14.9	15.0	15.1	15.2	15.3	15.4	15.5	15.6	15.7	15.8	15.9	16.0	16.1	16.2	16.3	16.4	16.5	16.7
17.5	14.6	14.7	14.8	14.9	15.0	15.1	15.2	15.3	15.4	15.5	15.6	15.7	15.8	15.9	16.0	16.1	16.2	16.3	16.4	16.5
18.0	14.4	14.6	14.7	14.8	14.9	15.0	15.1	15.2	15.3	15.4	15.5	15.6	15.7	15.8	15.9	16.0	16.1	16.2	16.3	16.4
18.5	14.3	14.4	14.5	14.6	14.7	14.8	15.0	15.1	15.2	15.3	15.4	15.5	15.6	15.7	15.8	15.9	16.0	16.1	16.2	16.3

(*A*)	14.0	14.1	14.2	14.3	14.4	14.5	14.6	14.7	14.8	14.9	15.0	15.1	15.2	15.3	15.4	15.5	15.6	15.7	15.8	15.9
(*T*)																				
19.0	14.2	14.3	14.4	14.5	14.6	14.7	14.8	14.9	15.0	15.1	15.2	15.3	15.4	15.5	15.6	15.7	15.8	16.0	16.1	16.2
19.5	14.1	14.2	14.3	14.4	14.5	14.6	14.7	14.8	14.9	15.0	15.1	15.2	15.3	15.4	15.5	15.6	15.7	15.8	15.9	16.0
20.0	14.0	14.1	14.2	14.3	14.4	14.5	14.6	14.7	14.8	14.9	15.0	15.1	15.2	15.3	15.4	15.5	15.6	15.7	15.8	15.9
20.5	13.9	14.0	14.1	14.2	14.3	14.4	14.5	14.6	14.7	14.8	14.9	15.0	15.1	15.2	15.3	15.4	15.5	15.6	15.7	15.8
21.0	13.8	13.9	14.0	14.1	14.2	14.3	14.4	14.5	14.6	14.7	14.8	14.9	15.0	15.1	15.2	15.3	15.3	15.4	15.5	15.6
21.5	13.7	13.8	13.8	13.9	14.0	14.1	14.2	14.3	14.4	14.5	14.6	14.7	14.8	14.9	15.0	15.1	15.2	15.3	15.4	15.5
22.0	13.5	13.6	13.7	13.8	13.9	14.0	14.1	14.2	14.3	14.4	14.5	14.6	14.7	14.8	14.9	15.0	15.1	15.2	15.3	15.4
22.5	13.4	13.5	13.6	13.7	13.8	13.9	14.0	14.1	14.2	14.3	14.4	14.5	14.6	14.7	14.8	14.9	15.0	15.1	15.2	15.3
23.0	13.3	13.4	13.5	13.6	13.7	13.8	13.9	14.0	14.1	14.2	14.3	14.4	14.5	14.6	14.6	14.7	14.8	14.9	15.0	15.1
23.5	13.2	13.3	13.4	13.5	13.6	13.7	13.8	13.8	13.9	14.0	14.1	14.2	14.3	14.4	14.5	14.6	14.7	14.8	14.9	15.0
24.0	13.1	13.2	13.2	13.3	13.4	13.5	13.6	13.7	13.8	13.9	14.0	14.1	14.2	14.3	14.4	14.5	14.6	14.7	14.8	14.9
24.5	12.9	13.0	13.1	13.2	13.3	13.4	13.5	13.6	13.7	13.8	13.9	14.0	14.1	14.2	14.3	14.4	14.4	14.5	14.6	14.7
25.0	12.8	12.9	13.0	13.1	13.2	13.3	13.4	13.5	13.6	13.7	13.8	13.8	13.9	14.0	14.1	14.2	14.3	14.4	14.5	14.6
25.5	12.7	12.8	12.9	13.0	13.1	13.2	13.3	13.3	13.4	13.5	13.6	13.7	13.8	13.9	14.0	14.1	14.2	14.3	14.4	14.5
26.0	12.6	12.7	12.8	12.8	12.9	13.0	13.1	13.2	13.3	13.4	13.5	13.6	13.7	13.8	13.9	14.0	14.1	14.1	14.2	14.3
26.5	12.4	12.5	12.6	12.7	12.8	12.9	13.0	13.1	13.2	13.3	13.4	13.5	13.6	13.6	13.7	13.8	13.9	14.0	14.1	14.2
27.0	12.3	12.4	12.5	12.6	12.7	12.8	12.9	13.0	13.1	13.1	13.2	13.3	13.4	13.5	13.6	13.7	13.8	13.9	14.0	14.1
27.5	12.2	12.3	12.4	12.5	12.6	12.7	12.7	12.8	12.9	13.0	13.1	13.2	13.3	13.4	13.5	13.6	13.7	13.7	13.8	13.9
28.0	12.1	12.2	12.2	12.3	12.4	12.5	12.6	12.7	12.8	12.9	13.0	13.1	13.2	13.2	13.3	13.4	13.5	13.6	13.7	13.8
28.5	11.9	12.0	12.1	12.2	12.3	12.4	12.5	12.6	12.7	12.8	12.8	12.9	13.0	13.1	13.2	13.3	13.4	13.5	13.6	13.7
29.0	11.8	11.9	12.0	12.1	12.2	12.3	12.4	12.4	12.5	12.6	12.7	12.8	12.9	13.0	13.1	13.2	13.3	13.3	13.4	13.5
29.5	11.7	11.8	11.9	12.0	12.0	12.1	12.2	12.3	12.4	12.5	12.6	12.7	12.8	12.8	12.9	13.0	13.1	13.2	13.3	13.4
30.0	11.6	11.6	11.7	11.8	11.9	12.0	12.1	12.2	12.3	12.4	12.4	12.5	12.6	12.7	12.8	12.9	13.0	13.1	13.2	13.2

4.3 Total Acidity

Total acidity is determined by titration with a titrated alkaline solution until the free carboxyl groups are neutralized. Neutralization is perceived either by using an indicator that changes color to a specific pH or by using a pH meter. CO_2 and SO_2 should not be included in total acidity.

4.3.1 Titration with Indicator (Bromothymol Blue)

Principle
This technique is based on the change in the color of the bromothymol blue indicator that we added to the must or wine when, upon the addition of the alkaline solution, the pH acquires the value 7.

Apparatus

- Conical flask Erlenmeyer, capacity 250 mL
- Vacuum pump or water pump with water
- Beaker 50 mL
- Burette
- Volumetric cylinder
- Pipette 5 mL
- Must or wine 30 mL
- Distilled water
- Bromothymol blue solution 4‰
- Solution NaOH *N*/10

Procedure
With strong stirring under vacuum for 3 min, we remove the CO_2 present in the must or wine.

In a boiling flask, we put 30 mL of boiled distilled water, 1 mL of bromothymol blue solution 4‰ and 5 mL of must or wine without CO_2. Stir the contents of the beaker well and then titrate with *N*/10 NaOH solution until the characteristic blue-green color appears and is maintained for at least 15 s. We note the mL of *N*/10 NaOH solution consumed to neutralize the acids and then do our calculations to express the total acidity.

Calculation
To calculate the total acidity in meq/L, we use the following formula:

$$\text{T.A} = \frac{n}{v} \times N \times 1000 \text{ meq / L} \quad (4.1)$$

n = the amount of NAOH solution, $N/10$ consumed in mL.
v = the amount in mL of the must or wine sample we used.
N = the normality of the NaOH solution.

To express the total acidity in g/L of sulfuric or tartaric acid, all we have to do is multiply the meq we found by the weight of meq of the corresponding acid. For sulfuric acid, it is 0.049 g, and for tartaric acid, it is 0.075 g.

4.3.2 Titration with pH Meter

We repeat the previous method with the difference that the entire time of the titration we have the sensor of our pH meter immersed in the sample to be measured and we stop the titration when our pH meter shows the number 7. At this point, we note the consumption of the μL of the NAOH solution $N/10$ and by applying the same formula (4.1), we express the total acidity.

4.4 pH

Principle
The potential difference between two electrodes submerged in a test liquid is measured. One electrode's potential varies with the pH of the liquid, while the other maintains a constant, known potential and serves as the reference electrode.

Apparatus

- pH meter with a scale calibrated in pH units and enabling measurements to be made to at least ±0.01 pH units.
- Electrodes:
  - a glass electrode, stored in distilled water;
  - a calomel-saturated potassium chloride reference electrode, preserved in a saturated potassium chloride solution; or,
  - a combined electrode, maintained in distilled water.

Reagents

- Buffer for pH (3.57 at 20 °C, 3.56 at 25 °C, 3.55 at 30 °C). Saturated potassium hydrogen tartrate solution, containing 5.7 g/L potassium hydrogen tartrate ($CO_2HC_2H_4O_2CO_2K$) at 20 °C.
- Buffer for pH (4.003 at 20 °C, 4.008 at 25 °C, 4.015 at 30 °C). Potassium hydrogen phthalate solution, 0.05 M, containing 10.211 g/L potassium hydrogen phthalate, $CO_2HC_6H_4CO_2K$, at 20 °C.

- Buffer for pH (6.88 at 20 °C, 6.86 at 25 °C, 6.85 at 30 °C). Add 3.402 g of potassium di-hydrogen phosphate, KH_2PO_4, 4.354 g di-potassium hydrogen phosphate, K_2HPO_4 and fill with distilled water to 1 L.

Procedure
Calibration of the pH meter should be performed at 20 °C with standard buffer solutions. The selected pH values should cover the entire spectrum likely to be found in musts and wines.

The electrode must wash up with distilled water before any measure. Dip the electrode into the sample to be analyzed, the temperature of which should be between 20 and 25 °C and as close as possible to 20 °C. Read the pH value directly from the scale. Perform at least two measurements on the same sample. The final result should be the arithmetic mean of the two measurements.

Calculation: Expression of Results
The pH meter reading expressed with two decimal places exactly as it appears on the screen constitutes the pH of the wine being measured.

4.5 Volatile Acidity

For the determination of volatile acidity apart from the automatic analyzers now widely used in the wine industry, two methods are applied. The first is the method of simple distillation of the wine and the second is the steam distillation of wine.

4.5.1 Determination of Volatile Acidity with Simple Distillation

Principle
These methods involve collecting volatile acids through distillation and titrating them with an alkaline solution using phenolphthalein as an indicator. To facilitate the distillation of the wine, it is acidified by adding crystalline tartaric acid (0.5 g per 20 mL of wine), which helps to liberate the volatile acids from their salts into the distillate. It is important to note that volatile acidity measurements should exclude acidity from carbonic acid, both free and bound sulfite anhydride, and sorbic acid.

Apparatus

- Simple distilling device
- Conical flask of 200 mL
- Pipette 10, 20, and 25 mL
- Vacuum pump or water pump with water
- Burette
- Beakers

Reagents

- Wine 50 mL
- Crystalline tartaric acid
- Solution NaOH *N*/10
- Iodine solution *N*/100
- Phenolphthalein
- Starch solution, 5 g/L*

* **How to make starch solution.**

Mix 5 g of starch with about 500 mL of water. Bring to the boil, stirring continuously and boil for 10 min. Add 200 g sodium chloride. When cool, make up to 1 L.

- Saturated solution of sodium tetraborate, $Na_2B_4O_7 \bullet 10H_2O$, about 55 g/L at 20 °C. (Borax)

Procedure

- *Preparation of Wine Sample*
  Place about 50 mL of wine in a vacuum flask and apply vacuum to the flask with the water pump for 1–2 min while shaking continuously. Other CO_2 elimination systems may be used if the CO_2 elimination is guaranteed.
- *Simple Distillation*
  In the 200 mL spherical flask that is part of the still, we put 20 mL of decarbonated wine, 35 mL of distilled water, and 5 g of crystal tartaric acid. Then we start the distillation with a constant flame and collect exactly 50 mL of the mixture. The distillate is then transferred to a beaker where the distilled wash water of the volumetric flask is added.
- *Titration with NaOH N/10 (n)*
  The mixture contained in the beaker is titrated with a *N*/10 NaOH solution in the presence of phenolphthalein 1% indicator. When the solution acquires the characteristic pink shade which remains for a few seconds, stop the titration and note the indication (n) of the burette.
- *First Titration with Iodine Solution (I_2) N/100 (n′)*
  Then to correct the effect of sulfuric anhydride on the titrated solution, we add one drop of hydrochloric acid and 2 mL of starch solution and start a second titration this time with *N*/100 iodine solution. When the solution acquires the characteristic blue shade, we stop the titration and note the indication (n') of the consumed iodine.
- *Second Titration with Iodine Solution (I2) N/100 (n″)*
  Then we add a few drops of saturated sodium tetraborate solution (Borax) solution until the solution acquires a pink hue. We again titrate the binary solution with iodine *N*/100 until the characteristic blue color and note the consumption of (n'').

Expression of Results

The volatile acidity, expressed in milliequivalents per liter to one decimal place, is given by

$$\mathrm{AV} = 5 \times \left(n - \frac{n}{10} - \frac{n''}{20} \right) \times 1.25$$

The volatile acidity, expressed in grams of sulfuric acid per liter to two decimal places, is given by

$$\mathrm{AV} = \left(n - \frac{n'}{10} - \frac{n''}{20} \right) \times 0.306$$

The volatile acidity, expressed in grams of acetic acid per liter to two decimal places, is given by

$$\mathrm{AV} = \left(n - \frac{n'}{10} - \frac{n''}{20} \right) \times 0.375$$

4.5.2 Determination of Volatile Acidity with Steam Distillation

Principle

These methods involve collecting volatile acids through distillation and titrating them with an alkaline solution using phenolphthalein as an indicator. To facilitate the distillation of the wine, it is acidified by adding crystalline tartaric acid (0.5 g per 20 mL of wine), which helps to liberate the volatile acids from their salts into the distillate. It is important to note that volatile acidity measurements should exclude acidity from carbonic acid, both free and bound sulfite anhydride, and sorbic acid [1].

Apparatus

- **Steam distillation** apparatus consisting of:
  - a steam generator; the steam must be free of carbon dioxide
  - a flask with steam pipe
  - a distillation column
  - a condenser

This equipment must pass the following three tests:

(a) Add 20 mL of boiled water to the flask. Collect 250 mL of the distillate and mix in 0.1 mL of 0.1 M sodium hydroxide solution and two drops of phenolphthalein solution. Ensure the pink color remains stable for at least 10 s, indicating the steam is free of carbon dioxide.

(b) Add 20 mL of 0.1 M acetic acid solution into the flask. Gather 250 mL of the distillate. Proceed to titrate with 0.1 M sodium hydroxide solution; the titer volume should be no less than 19.9 mL, confirming that at least 99.5% of the acetic acid has been carried over with the steam.

(c) Add 20 mL of 1 M lactic acid solution into the flask. Collect 250 mL of the distillate and titrate the acid using 0.1 M sodium hydroxide solution.

The volume of sodium hydroxide solution added must be less than or equal to 1.0 mL (i.e., not more than 0.5% of lactic acid is distilled).

Any apparatus or procedure which passes these tests satisfactorily fulfills the requirements of official international apparatus or procedures.

- Water aspirator vacuum pump
- Vacuum flask
- Conical flask of 250 mL
- Pipette of 20 mL
- Burette
- Beakers

Reagents

Tartaric acid, crystalline

- Sodium hydroxide solution, 0.1 M
- Phenolphthalein solution, 1%, in neutral alcohol, 96% (m/v)
- Iodine solution, 0.005 M
- Starch solution, 5 g/L*

* **How to make Starch solution.**

Mix 5 g of starch with about 500 mL of water. Bring to the boil, stirring continuously and boil for 10 min. Add 200 g sodium chloride. When cool, make up to 1 L.

- Saturated solution of sodium tetraborate, $Na_2B_4O_7 \bullet 10H_2O$, about 55 g/L at 20 °C (Borax).
- Acetic acid, 0.1 M
- Lactic acid solution, 0.1 M**

** **How to make Lactic acid solution.**

Dilute 100 mL of lactic acid in 400 mL of water. Heat this solution in an evaporating dish over a boiling water bath for 4 h, replenishing the volume as needed with distilled water. Once cooled, bring the total volume up to 1 L. Titrate 10 mL of this solution with a 1 M sodium hydroxide solution. Adjust the solution to achieve a 1 M concentration of lactic acid (90 g/L).

Procedure

- *Preparation of Sample*

  To eliminate CO_2 from wine sample, place about 50 mL of wine in a vacuum flask, apply vacuum to the flask with the water pump for 1–2 min while shaking continuously. Other CO_2 elimination systems may be used if the CO_2 elimination is guaranteed.

- *Steam Distillation*
  Place 20 mL of wine, freed from carbon dioxide into the flask. Add about 0.5 g of crystalline tartaric acid. Collect at least 250 mL of the distillate.
- *First Titration (n)*
  Titrate with the 0.1 M sodium hydroxide solution, using few drops of phenolphthalein as indicator. Note down as (n) mL be the volume of sodium hydroxide used.
- *Second Titration (n')*
  After the first titration, we add 3–4 drops of the dilute hydrochloric acid, 2 mL starch solution and a few crystals of potassium iodide. Titrate the free sulfur dioxide with 0.005 M iodine solution. Note down as (n') mL be the volume used.
- *Third Titration (n")*
  After the second titration, we add the saturated sodium tetraborate solution until the pink coloration appears again. Then titrate the combined sulfur dioxide with the 0.005 M iodine solution and note down as (n'') mL be the volume used.

Expression of Results

The volatile acidity, expressed in milliequivalents per liter to one decimal place, is given by the following type:

$$\mathrm{VA} = \left(n - \frac{n'}{10} - \frac{n''}{20}\right) \times 5$$

The volatile acidity, expressed in grams of sulfuric acid per liter to two decimal places is given by the following type:

$$\mathrm{VA} = \left(n - \frac{n'}{10} - \frac{n''}{20}\right) \times 0.245$$

The volatile acidity, expressed in grams of acetic acid per liter to two decimal places, is given by the following type:

$$\mathrm{VA} = \left(n - \frac{n'}{10} - \frac{n''}{20}\right) \times 0.300$$

4.6 Tartaric Acid

4.6.1 Method 1

Principle

Tartaric acid is precipitated in the form of acid potassium tartrate under such conditions that the solubility of the tartrate is reduced to a minimum. An oximetric determination is then performed.

Apparatus

- Centrifugal
- Conical flask of 200 mL
- Pipettes of 10 mL
- Rod made of glass
- Fridge
- Burette of 50 mL
- Beakers

Reagents

- Precipitation solution
  - Potassium chloride p.a. (300 g)
  - Potassium sodium tartrate tetrahydrate (Seignette salt) (5.75 g)
  - Potassium oxalate (1.25 g)
  - Distilled water (1000 mL)
- Ethyl alcohol 96% vol
- Flushing solution
  - Potassium chloride (150 g)
  - Ethyl alcohol 96% vol. (200 mL)
  - Distilled water to 1000 mL
- Bromothymol Blue Indicator
  - Bromothymol Blue (0.5 g)
  - Ethyl alcohol 50% vol. (50 mL)
- Sodium Hydroxide *N*/10
- Acetic acid

Procedure

First, in a tube of our centrifuge, we place 10 mL of the wine we want to examine. There we add 0.4 mL of acetic acid and 2 mL of ethyl alcohol and 7.5 mL of our dipping solution. This is done in order to convert the tartaric acid into a stable sediment. The contents of the tube are stirred steadily with the rod made of glass until a crystalline cloud is seen, this is done in about 15 s of stirring.

The tube will then be placed in the refrigerator at 8 °C for 70 min and stirred again with the glass rod. Then the tube is placed again for 70 min in the refrigerator at the same temperature.

Then place the tube in the centrifuge for 10 min at 700–800 rpm. At the end of the centrifugation, the sediment is immediately separated from the liquid in the tube. This is done in two ways either by carefully decanting the supernatant from the tube or by filtration on a no. 4 pore porcelain filter.

The sediment is washed two times with 2 mL of flushing solution. After each washing, centrifugation and separation of the liquid from the solid sediment is repeated.

After the sediment has now been separated, it is transferred with distilled water to a boiling glass, followed by titration of the tartaric acid with sodium hydroxide solution *N*/10 in the presence of bromothymol blue indicator (2–3 drops). Titration stops when the initial yellow color turns blue. Note as (n) the consumption of sodium hydroxide.

Calculation: Expression of Results

The amount of tartaric acid in the wine sample we examined depending on the way we want to express our result is equal to:

$$\text{T.A} = 13.33 \times (1.5n - 2)\,\text{meq / L}$$
$$\text{T.A} = 1.254 \times (1.5n - 2)\,\text{g / L potassium tartrate}$$
$$\text{T.A} = (1.5n - 2)\,\text{g / L of tartaric acid}$$

4.6.2 Gravimetric Method [2–5]

Principle

Tartaric acid precipitates as calcium (±)tartrate and is measured gravimetrically. This measurement can also be performed using a volumetric method for comparison. The precipitation conditions (pH, total volume, and concentrations of precipitating ions) ensure complete precipitation of calcium (±)tartrate, while calcium D(−) tartrate stays in solution.

When meta-tartaric acid has been added to the wine, which causes the precipitation of the calcium (±)tartrate to be incomplete, it must first be hydrolyzed.

Apparatus

- High accuracy weight scale
- 600 mL beaker
- Glass rod
- Vacuum flask
- Sintered glass crucible of porosity No. 4
- Oven
- 50 mL conical flask
- Reflux condenser
- Stoppered bottle

Reagents

- Calcium acetate solution containing 10 g of calcium per liter:
  - Calcium carbonate, $CaCO_3$ (25 g)
  - Acetic acid, glacial, CH_3COOH (= 1.05 g/mL) (40 mL)
  - Distilled water to 1000 mL
- Calcium (±)tartrate, crystallized: $CaC_4O_6H_4 \bullet 4H_2O$

- Place 20 mL of L(+) tartaric acid solution, 5 g/L, into a 400 mL beaker. Add 20 mL of ammonium D(−) tartrate solution, 6.126 g/L, and 6 mL of calcium acetate solution containing 10 g of calcium per liter. Let it stand for 2 h to allow precipitation. Collect the precipitate in a sintered glass crucible of porosity No. 4, and wash it three times with about 30 mL of distilled water. Then dry to constant weight in the oven at 70 °C. Using the quantities of reagent indicated above, about 340 mg of crystallized calcium (±) tartrate is obtained. Store in a stoppered bottle.
- Precipitation solution (pH 4.75)
  - D(−) ammonium tartrate (150 mg)
  - Calcium acetate solution, 10 g calcium/L (8.8 mL)
  - Distilled water to (1000 mL)

Dissolve the D(−) ammonium tartrate in 900 mL water; add 8.8 mL calcium acetate solution and make up to 1000 mL. Since calcium (±) tartrate is slightly soluble in this solution, add 5 mg of calcium (±) tartrate per liter, stir for 12 h and filter.

We can make the precipitation solution from D(−) tartaric acid as above.

- Alternative precipitation solution
  - D(−) tartaric acid (122 mg)
  - Ammonium hydroxide solution (= 0.97 g/mL), 25% (v/v) (0.3 mL)

Instructions to Make Ammonium Hydroxide Solution

Dissolve the D(−) tartaric acid, add the ammonium hydroxide solution, and make up to about 900 mL, then add 8.8 mL of calcium acetate solution, make up to a liter and adjust the pH to 4.75 with addition of acetic acid. Since calcium (±) tartrate is slightly soluble in this solution, add 5 mg of calcium (±) tartrate per liter, and stir for 12 h and then filter.

Dissolve the D(−) tartaric acid, add the ammonium hydroxide solution, and make up to about 900 mL, then add 8.8 mL of calcium acetate solution, make up to a liter and adjust the pH to 4.75 with addition of acetic acid. Since calcium (±) tartrate is slightly soluble in this solution, add 5 mg of calcium (±) tartrate per liter, and stir for 12 h and then filter.

Procedure

- *Wines Without Addition of Meta-Tartaric Acid*
- Place 500 mL of precipitation solution and 10 mL of wine into a 600-mL beaker. Mix and initiate precipitation by rubbing the sides of the vessel with the tip of a glass rod. Leave to precipitate for 12 h. Filter the liquid and precipitate through a weighed sintered glass crucible of porosity No. 4 fitted on a clean vacuum flask. Rinse the vessel in which precipitation took place with the filtrate to ensure that all precipitate is transferred.

- Dry to constant weight in an oven at 70 °C. Weigh. Note as (p) be the weight of crystallized calcium (±) tartrate, $CaC_4O_6H_4 \bullet 4H_2O$, obtained.
- *Wines with Addition of Meta-Tartaric Acid*
- When analyzing wines to which meta-tartaric acid has been or is suspected of having been added, proceed by first hydrolyzing this acid as follows:
- Put 10 mL of wine and 0.4 mL of glacial acetic acid, CH_3COOH, ($\rho_{20} = 1.05$ g/mL) into a 50-mL conical flask. Place a reflux condenser on top of the flask and boil for 30 min. Allow to cool and then transfer the solution in the conical flask to a 600 mL beaker. Rinse the flask twice using 5 mL of water each time and then continue as described above. Meta-tartaric acid is calculated and included as tartaric acid in the final result.

Calculation: Expression of Results

One molecule of calcium (±)tartrate corresponds to half a molecule of L(+) tartaric acid in the wine.

The quantity of tartaric acid per liter of wine, expressed in milliequivalents, is equal to:

$$\mathrm{T.A} = 384.5 \times p$$

It is quoted to one decimal place.

The quantity of tartaric acid per liter of wine, expressed in grams of tartaric acid, is equal to:

$$\mathrm{T.A} = 28.84 \times p$$

It is quoted to one decimal place.

The quantity of tartaric acid per liter of wine, expressed in grams of potassium tartrate, is equal to:

$$\mathrm{T.A} = 36.15 \times p$$

It is quoted to one decimal place.

4.6.3 Comparative Volumetric Analysis [3, 6, 7]

Apparatus

- Vacuum flask
- Pipettes
- Burette of 50 mL

Reagents

- Hydrochloric acid ($\rho_{20} = 1.18$–1.19 g/mL) diluted 1:5 with distilled water
- EDTA solution, 0.05 M

  - EDTA (ethylenediaminetetraacetic acid disodium salt) (18.61 g)
  - Water to 1000 mL
- Sodium hydroxide solution, 40% (m/v)
  - Sodium hydroxide, NaOH (40 g)
  - Water to 100 mL
- Complexometric indicator: 1% (m/m)
  - 2-hydroxy-1-(2-hydroxy-4-sulpho-1-naphthylazo)-3-naphthoic acid (1 g)
  - Sodium sulfate, Na_2SO_4 (anhydrous) (100 g)

Procedure

After weighing, replace the sintered glass crucible containing the precipitate of calcium (±) tartrate on the vacuum flask and dissolve the precipitate with 10 mL of dilute hydrochloric acid. Wash the sintered glass crucible with 50 mL of distilled water. Add 5 mL 40% sodium hydroxide solution and about 30 mg of indicator. Titrate with EDTA solution, 0.05 M. Let the number of mL used be (*n*).

Calculation: Expression of Results

The quantity of tartaric acid per liter of wine, expressed in milliequivalents, is equal to:

$$\mathrm{T.A} = 5 \times n$$

It is quoted to one decimal place.

The quantity of tartaric acid per liter of wine, expressed in grams of tartaric acid, is equal to:

$$\mathrm{T.A} = 0.375 \times n$$

It is quoted to one decimal place.

The quantity of tartaric acid per liter of wine, expressed in grams of potassium acid tartrate, is equal to:

$$\mathrm{T.A} = 0.470 \times n$$

It is quoted to one decimal place.

4.7 Malic Acid [8]

Principle

Malic acid is isolated using an anion exchange column and quantified colorimetrically in the eluate by assessing the yellow hue it produces with chromotropic acid in the presence of concentrated sulfuric acid. To correct for interfering substances, the absorbance measured with 86% sulfuric and chromotropic acid (where malic acid is non-reactive) is subtracted from the absorbance measured with 96% strength acids.

Apparatus

- Spectrophotometer set to measure absorbance at 420 nm using cells of 1-cm optical path.
- Glass column 250 mm approximately in length and 35 mm internal diameter, fitted with drain tap.
- Glass column approximately 300 mm in length and 10–11 mm internal diameter, fitted with drain tap.
- Thermostatically controlled water bath at 100 °C.
- Wide necked 30-mL tubes fitted with ground glass stoppers.
- 50- and 100-mL volumetric flasks.
- Pipettes 5, 10, 15, and 20 mL.

Reagents

- A strongly basic anion exchanger (e.g. Merck III)
- Sodium hydroxide, 5% (m/v)
- Acetic acid, 0.5% (m/v)
- Acetic acid, 30% (m/v)
- Sodium sulfate solution, 10% (m/v)
- Sulfuric acid, 86% (m/m)
- Concentrated sulfuric acid, 95–97% (m/m)
- Chromotropic acid, 5% (m/v)

Prepare fresh solution before each determination by dissolving 500-mg sodium chromotropate, $C_{10}H_6Na_2O_8S_2 \bullet 2H_2O$, in 10-mL distilled water.

- 0.5 g DL-malic acid per liter solution

Dissolve 250 g malic acid ($C_4H_6O_5$) in sodium sulfate solution, 10%, to obtain 500 mL.

Procedure

- *Preparation of Ion Exchanger*
  In a 35 × 250 mm glass column, we place a plug of cotton impregnated with distilled water. Introduce a suspension of anion exchange resin into the column, ensuring the liquid level is 50 mm above the resin's surface. Rinse with 1000 mL of distilled water. Proceed to wash the column with a 5% sodium hydroxide solution, allowing it to drain until it's 2–3 mm from the resin's top, and perform two additional washes, leaving it for 1 h after the last. Rinse again with 1000 mL of distilled water. Fill the column with a 30% acetic acid solution, drain to 2–3 mm above the resin, and conduct two more acetic acid washes. Let the column sit for

at least 24 h before use. Store the ion exchange resin in 30% acetic acid for future analysis.

- *Preparation of Ion Exchange Column*
  Place a plug of cotton wool at the bottom of the column measuring 11 × 300 mm above the tap. Add the prepared ion exchanger up to a 10 cm height. Open the tap to let the 30% acetic acid solution drain until it's about 2–3 mm above the ion exchanger's surface. Then, rinse the exchanger with 50 mL of 0.5% acetic acid solution.
- *Separation of* dl-*Malic Acid*
  Pour onto the column 10 mL of wine or must. Allow to drain drop by drop (average rate of one drop per second) and stop the flow 2–3 mm from the top of the resin. Rinse the column with 50 mL of 0.5% (m/v) acetic acid, followed by 50 mL of distilled water. Allow it to drain at the same rate as before, halting the flow when it is 2–3 mm from the resin's surface. Elute the acids absorbed on the exchange resin with sodium sulfate solution, 10%, at the same rate as in the previous steps (1 drop/s). Gather the eluate in a 100-mL volumetric flask. The ion exchange column may be regenerated by following the procedure outlined in the initial step.
- *Determination of Malic Acid*
  Take two wide necked 30-mL tubes fitted with ground glass stoppers, A and B. In each tube, add 1.0 mL of the eluate and 1.0-mL chromotropic acid solution, 5%. Add to tube A 10.0 mL sulfuric acid, 86% (m/m), (reference) and to the tube B 10.0 mL sulfuric acid, 96% (m/m), (sample). Secure the stopper and gently shake to homogenize, ensuring the glass stopper remains dry. Immerse the tubes in a boiling water bath for exactly 10 min. Cool the tubes in darkness at 20 °C for precisely 90 min. Promptly measure the absorbance of tube B in comparison to the sample in tube A at a wavelength of 420 nm using 1 cm cuvettes.
- *Plotting the Calibration Curve*
  Pipette 5, 10, 15, and 20 mL of the DL-malic acid solution (0.5 g/L) into separate 50-mL volumetric flasks. Adjust the volume to the mark with a 10% sodium sulfate solution. These solutions correspond to eluates obtained from wines containing 0.5, 1.0, 1.5, and 2.0 g DL-malic acid per liter. Continue as indicated in the previous step. The graph depicting the absorbance of these solutions against their malic acid concentration is expected to be a straight line that passes through the origin. The color intensity is largely dependent on the concentration of sulfuric acid utilized. It is necessary to check the calibration curve to see if the concentration of the sulfuric acid has changed.

Calculation: Expression of Results

To determine the concentration of DL-malic acid in grams per liter, plot the absorbance values on a calibration graph. The content should be expressed to one decimal place.

4.8 L-Ascorbic Acid [9]

Principle

Ascorbic acid undergoes conversion to dehydroascorbic acid on activated carbon. Subsequently, dehydroascorbic acid reacts with orthophenylenediamine (OPDA) to form a fluorescent compound. A control prepared in the presence of boric acid enables spurious fluorescence to be determined (by the formation of a boric acid/dehydroascorbic acid complex). The sample and the control are analyzed fluorometrically and the concentration of dehydroascorbic acid calculated.

Apparatus

- Fluorometer
  A spectrofluorometer equipped with a lamp giving a continuous spectrum and using it at minimum power. Determining the optimal excitation and emission wavelengths for the test is essential and varies based on the equipment utilized. The excitation wavelength will be approximately 350 nm and the emission wavelength approximately 430 nm.
- Sintered glass filter of porosity 3
- Test tubes (diameter approximately 10 mm)
- Filter papers
- Pipette of 2, 4, 5, 6 mL
- 100-mL graduated flasks
- Stirring rods for test tubes

Reagents

- Sodium acetate trihydrate solution ($CH_3COONa \bullet 3H_2O$), 500 g/L
- Orthophenylenediamine dihydrochloride solution ($C_6H_{10}C_{12}N_2$), 0.02% (m/v), prepared just before use
- Mixed solution of boric acid and sodium acetate:

Instructions to Make Boric Acid and Sodium Acetate Solution

Dissolve 3 g of boric acid (H_3BO_3) in 100 mL of a 500 g/L sodium acetate solution. This solution must be prepared just before use.

- Acetic acid solution (CH_3COOH) 56%: glacial acetic acid (ρ_{20} = 1.05 g/mL), diluted to 56% (v/v), pH approximately 1.2.
- L-Ascorbic acid standard solution, 1 g/L.
- Just before use, dissolve 50 mg of L-ascorbic acid previously dehydrated in a desiccator and protected against light, in 50 mL of acetic acid solution.
- Very pure analytical grade activated carbon.

Instructions to Make Very Pure Analytical Grade Activated Carbon
Place 100 g of activated carbon into a 2-L conical flask and add 500 -mL aqueous hydrochloric acid solution, 10% (v/v), (ρ_{20} = 1.19 g/mL). Heat until boiling, then strain through a sintered glass filter with a porosity of 3. Gather the carbon processed in this manner into a 2-L conical flask. Add 1 L of water, shake well, and then filter through a sintered glass filter with a porosity of 3. Perform this operation two additional times. Place the residue in an oven set to 115 ± 5 °C for 12 h.

Procedure

- *Preparation of the Sample of Wine or Must*
  Measure a volume of the wine or must and dilute it to 100 mL in a graduated flask using a 56% acetic acid solution to achieve an ascorbic acid concentration ranging from 0 to 60 mg/L. Shake the flask vigorously to thoroughly mix its contents. Add 2 g of activated carbon and allow to stand for 15 min, shaking occasionally. Filter the solution using standard filter paper, and discard the initial milliliters of the filtrate. Transfer 5 mL of the filtrate into each of two 100 mL graduated flasks. To the first, add 5 mL of the boric acid and sodium acetate mixture (sample blank), and to the second, add 5 mL of the sodium acetate solution (sample). Allow to stand for 15 min, stirring occasionally. Make to 100 mL with distilled water. Transfer 2 mL from the contents of each flask into a test tube and then add 5 mL of ortho-phenylenediamine solution. Stir with the stirring rod and allow the reaction proceed for 30 min in the dark and then make the spectrofluorometric measurements.
- *Preparation of the Calibration Curve*
  Pipette 2, 4, and 6 mL of the standard ascorbic acid solution into three separate 100-mL graduated flasks, respectively. Dilute each to 100 mL with acetic acid solution and mix thoroughly by stirring. The standard solutions prepared in this way contain 2, 4, and 6 mg per 100 mL of L-ascorbic acid respectively. Add 2 g of activated carbon to each of the flasks and allow to stand for 15 min, stirring occasionally. Filter the solution using standard filter paper and discard the initial milliliters of the filtrate. Transfer 5 mL of each filtrate into three separate 100-mL graduated flasks, constituting the first series. Repeat the procedure to acquire a second set of three graduated flasks. Add 5 mL of the boric acid and sodium acetate mixed solution to each flask in the first series (for the blank test), and add 5 mL of the sodium acetate solution to each flask in the second series. Let stand for 15 min, stirring occasionally. Make up to 100 mL with distilled water. Extract 2 mL from the contents of each flask and mix with 5 mL of orthophenylenediamine solution. Stir the mixture and let the reaction occur for 30 min in darkness before conducting the spectrofluorometric measurements.
- *Fluorometric Determination*
  Calibrate the measurement scale to zero using the appropriate control test sample for each solution. Measure the fluorescence intensity for each solution across the

calibration range and for the solution under analysis. Plot the calibration curve, which should be a straight line passing through the origin. Determine the concentration (C) of ascorbic acid and dehydroascorbic acid in the analyzed solution from the graph.

Calculation: Expression of Results

The concentration of L-ascorbic acid and dehydroascorbic acid in wine, measured in milligrams per liter, is determined by the formula:

$$\text{L-ascorbic acid} = C \times F \text{ mg / L}$$

F represents the dilution factor.

4.9 Sucrose (Qualitative Method)

The addition of sucrose to grape must to increase the alcoholic strength is a common method but is not allowed except in very few cases. For this reason, it is important to initially perform a qualitative analysis for the presence or absence of sucrose.

4.9.1 Qualitative Method (Chromatographic Method)

Principle

First, we decolorize the wine under examination with lead acetate, magnesium oxide, and potassium permanganate to pH between 8 and 9. At a temperature of 100 °C, we treat the discolored wine with an acidic solution of diphenylamine. The compound formed with sucrose is extracted with chloroform which acquires a blue color.

Apparatus

- Conical flask of 100 mL
- Water bath
- Testing tubes
- Pipettes of 2 and 10 mL
- Calibrated pipettes
- Beakers

Reagents

- Crystalline neutral lead acetate
- Magnesium oxide in powder form
- Solution of 2‰ potassium permanganate
- Chloroform
- Diphenylamine solution

Instructions to Make Diphenylamine Solution
To 10 mL of a 10% solution of diphenylamine, in pure alcohol, add 20 mL of crystalline acetic acid and 70 mL of pure hydrochloric acid. It is important to check the purity of the solution. To do this, add 2 mL of distilled water to 2 mL and leave it in the water bath at a temperature of 100 °C. Then cool down the reagent we had in the water bath and extract 1 mL of chloroform. If the chloroform layer does not acquire a blue color, the reagent is considered pure.

Procedure
In 10 mL of water, dissolve 200-mg magnesium oxide powder and 200-mg neutral lead acetate. These remain for 5 min in a water bath at a temperature of 95 °C and we add 2 mL of discolored wine and diluted in such a way that the sugar content is less than 1%. To the clear supernatant liquid, add 5 drops (4–5 mL) of concentrated lead acetate solution until the formation of a precipitate stops. Then there is mixing and standing of the aqueous solution until the supernatant liquid becomes clear. When this is done add about 1 mL of saturated sodium sulfate solution to remove excess lead.

We add 0.5 mL of potassium permanganate solution and after remaining in the water bath for 10 min at a temperature of 95 °C, the liquid is rapidly cooled by immersion in cold water and followed by filtration. Then, 2 mL of filtrate is brought into a test tube and there we add 2 mL of the diphenylamine reagent. We place it again in the water bath for 5 min, followed by rapid cooling by immersion in cold water and addition of 1 mL of chloroform. This immediately extracts the colored compound formed.

Expression of Results
When we have dry wines without sucrose addition, an ash coloring develops with a very light blue tint, when we have dry wines that have added sucrose a clear blue color is formed. In the case of sweet wines where sucrose has not been added, the color becomes golden yellow. In the case that sucrose has been added, the color turns green.

4.10 Dry Extract [1, 10–15]

The **total dry extract**, also known as the total dry matter, comprises all substances that are non-volatile under certain physical conditions. These conditions should be such that the substances making up the extract are minimally altered during the testing process out.

The **sugar-free extract** represents the difference between the total dry extract and the total sugars. The reduced extract is calculated as the difference between the total dry extract and the sum of total sugars exceeding 1 g/L, potassium sulfate exceeding 1 g/L, any present mannitol, and any additional chemical substances that have been added to the wine.

The **residual extract** is the sugar-free extract less the fixed acidity expressed as tartaric acid.

Principle

The residues weight, resulting from a wine sample absorbed into filter paper and dried in an airstream at a pressure of 20–25 mmHg and a temperature of 70 °C, is determined.

Apparatus

- *Oven*
- The oven must consists of α cylindrical basin with internal diameter of 27 cm and height 6 cm made of aluminum with an aluminum lid, heated to 70 °C and regulated to 1 °C. Except the cylindrical basin, there should be a tube with internal diameter of 25 mm connecting the oven to a vacuum pump providing a flow rate of 50 L/h. The air which previously dried by bubbling through concentrated sulfuric acid is circulated in the oven by a fan in order to achieve quick homogenous reheating. The rate of airflow is regulated by a tap and is to be 30–40 L/h and the pressure in the oven is 25 mm of mercury.
- *Stainless Steel Dishes*
- Stainless steel dishes, 60 mm in internal diameter and 25 mm in height, come with appropriate lids. Each dish holds 4–4.5 g of filter paper, sliced into fluted strips of 22 mm length. The filter paper undergoes an initial wash with 2 g/L hydrochloric acid for 8 h, followed by five water rinses and is then air-dried.

Calibration of Apparatus and Method

1. Seal check for dish lids: A dish with dried filter paper and its lid secured, once cooled in a desiccator containing sulfuric acid, should not show a weight gain exceeding 1 mg/h when left in the laboratory.
2. Dryness degree verification: A pure sucrose solution, at a concentration of 100 g/L, should result in a dry extract of 100 g ± 1 g/L.
3. Lactic acid solution test: A pure lactic acid solution, at 10 g/L, should yield a dry extract of no less than 9.5 g/L.

Instructions to Make Lactic Acid Solution

Dilute 10 mL of lactic acid to about 100 mL with water. Heat this solution in a dish on a boiling water bath for 4 h, adding distilled water if the volume falls below approximately 50 mL. Then, bring the solution up to 1 L and titrate 10 mL of it with a 0.1 M alkali. Finally, adjust the lactic acid solution to a concentration of 10 g/L.

Procedure

First, we must weigh the dish so we place it with the filter paper in the oven for 1 h. Stop the vacuum pump and promptly place the lid on the dish upon opening the oven. Cool it in a desiccator and weigh to the nearest 0.1 mg; the mass of the dish and lid is noted as (p_o) g.

Then we have to weigh the sample. Pour 10 mL of the wine sample into the weighing dish. Ensure the sample is fully absorbed by the filter paper. Then, place the dish in the oven for 2 h. Stop the vacuum pump and promptly place the lid on the dish upon opening the oven. Cool it in a desiccator and weigh to the nearest 0.1 mg; the mass of the dish and lid is noted as (p) g.

Note: It is important to consider the sample weight when analyzing very sweet wines or musts.

Calculation: Expression of Results

The total dry extract is given by the following type:

$$\text{T.D.E} = (p - p_o) \times 100\,\text{g/L}$$

In very sweet wines or musts, the total dry extract is given by the following type:

$$\text{T.D.E} = (p - p_o) \times \frac{\rho_{20}}{P} \times 1000\,\text{g/L}$$

In type ρ_{20} is density of wine or must in g/mL.

In type P is mass of sample in grams.

Note: To calculate the total dry extract, consider the quantities of glucose and fructose (reducing sugars) and the quantity of saccharose separately as follows:

Sugar-free extract = Total dry extract – reducing sugars (glucose + fructose) – saccharose

In instances where the analytical method permits sugar inversion, apply the subsequent formula for calculation:

Sugar-free extract = Total dry extract – reducing sugars (glucose + fructose) – [(Sugars after inversion – Sugars before inversion) × 0.95]

Inversion describes the process by which a stereoisomer is transformed into compounds exhibiting reversed stereoisomerism. Specifically, sugar inversion refers to the process of splitting sucrose into fructose and glucose. This is achieved by maintaining acidified sugar solutions (100 mL of sugar solution plus 5 mL of concentrated hydrochloric acid) at a minimum of 50 °C for at least 15 min in a water bath, which is kept at 60 °C until the solution's temperature reaches 50 °C. The final solution exhibits laevorotation as a result of the fructose content, whereas the initial solution demonstrates dextrorotation due to the sucrose present (Tables 4.6 and 4.7).

Table 4.6 For the calculation of the total dry extract content (g/L)

Density to 2 decimals	Third decimal place									
	0	1	2	3	4	5	6	7	8	9
1.00	0.0	2.6	5.1	7.7	10.3	12.9	15.4	18.0	20.6	23.2
1.01	25.8	28.4	31.0	33.6	36.2	38.8	41.3	43.9	46.5	49.1
1.02	51.7	54.3	56.9	59.5	62.1	64.7	67.3	69.9	72.5	75.1
1.03	77.7	80.3	82.9	85.5	88.1	90.7	93.3	95.9	98.5	101.1
1.04	103.7	106.3	109.0	111.6	114.2	116.8	119.4	122.0	124.6	127.2
1.05	129.8	132.4	135.0	137.6	140.3	142.9	145.5	148.1	150.7	153.3
1.06	155.9	158.6	161.2	163.8	166.4	169.0	171.6	174.3	176.9	179.5
1.07	182.1	184.8	187.4	190.0	192.6	195.2	197.8	200.5	203.1	205.8
1.08	208.4	211.0	213.6	216.2	218.9	221.5	224.1	226.8	229.4	232.0
1.09	234.7	237.3	239.9	242.5	245.2	247.8	250.4	253.1	255.7	258.4
1.10	261.0	263.6	266.3	268.9	271.5	274.2	276.8	279.5	282.1	284.8
1.11	287.4	290.0	292.7	295.3	298.0	300.6	303.3	305.9	308.6	311.2
1.12	313.9	316.5	319.2	321.8	324.5	327.1	329.8	332.4	335.1	337.8
1.13	340.4	343.0	345.7	348.3	351.0	353.7	356.3	359.0	361.6	364.3
1.14	366.9	369.6	372.3	375.0	377.6	380.3	382.9	385.6	388.3	390.9
1.15	393.6	396.2	398.9	401.6	404.3	406.9	409.6	412.3	415.0	417.6
1.16	420.3	423.0	425.7	428.3	431.0	433.7	436.4	439.0	441.7	444.4
1.17	447.1	449.8	452.4	455.1	457.8	460.5	463.2	465.9	468.6	471.3
1.18	473.9	476.6	479.3	482.0	484.7	487.4	490.1	492.8	495.5	498.2
1.19	500.9	503.5	506.2	508.9	511.6	514.3	517.0	519.7	522.4	525.1
1.20	527.8	–	–	–	–	–	–	–	–	–

Table 4.7 Interpolation table

Fourth decimal place	Extract g/L	Fourth decimal place	Extract g/L	Fourth decimal place	Extract g/L
1	0.3	4	1.0	7	1.8
2	0.5	5	1.3	8	2.1
3	0.8	6	1.6	9	2.3

4.11 Ash

Ash content refers to all the products that remain after the wine residue is ignited post-evaporation. This process ensures that all cations, except for the ammonium cation, are transformed into carbonates or other anhydrous inorganic salts.

Principle

Wine undergoes ignition at temperatures ranging from 500 to 550 °C, ensuring the complete combustion (oxidation) of the organic material.

Apparatus

- Water bath at 100 °C
- Balance sensitive to 0.1 mg
- Temperature-controlled electric muffle furnace
- Dessicator
- Hot plate or infra-red evaporator
- Pipette of 20 mL
- Flat-bottomed platinum dish 70 mm in diameter and 25 mm in height

Procedure

Transfer with a pipette of 20 mL of wine into the previously tared platinum dish (original weight p_o g). Evaporate the substance on a boiling water bath, and then heat the residue on a hot plate at 200 °C or with an infra-red evaporator until carbonization starts. When no more fumes are produced, place the dish in the electric muffle furnace maintained at 525 ± 25 °C. After 15 min of carbonization, remove the dish from the furnace. Then, add 5 mL of distilled water, allow it to evaporate on the water bath or under the infrared evaporator, and subsequently heat the residue to 525 °C for 10 min. If the oxidization of carbonized particles is incomplete, the process must be repeated: washing the carbonized particles, evaporation of water, and ignition. After cooling in the desiccator, the dish is weighed (p_1 g).

The weight of the ash in the sample (20 mL) is then calculated as $p = (p_1 - p_o)$ g.

Note: In the case of wines with high sugar content, it is beneficial to add a few drops of pure vegetable oil to the extract prior to the initial ashing to prevent excessive foaming.

Calculation: Expression of Results

The weight P of the ash in grams per liter is given to two decimal places by the expression: $P = 50\,p$.

4.12 Alkalinity of Ash [16]

The alkalinity of ash in wine is determined by the total cations present, excluding the ammonium ion, which are combined with the wine's organic acids.

Principle

The ash is dissolved in a specified quantity of hot standardized acid solution, and the surplus is measured by titration with methyl orange as the indicator.

Apparatus

- Boiling water bath
- Pipette of 10 and 20 mL

Reagents

- Sulfuric acid solution, 0.05 M H_2SO_4
- Sodium hydroxide solution, 0.1 M NaOH
- Methyl orange, 0.1% solution in distilled water

Procedure

Add 10 mL of 0.05 M sulfuric acid solution to the ash obtained from 20 mL of wine in the platinum dish. Set the dish in the boiling water bath for approximately 15 min, stirring and breaking up the residue with a glass rod to expedite the dissolution process. Add two drops of methyl orange solution and titrate the excess sulfuric acid against 0.1 M sodium hydroxide until the color of the indicator changes to yellow.

Calculation: Expression of Results

The alkalinity of ash, expressed in milliequivalents per liter to one decimal place, is given by

$$A = 5 \times (10 - n)$$

where n mL is the volume of sodium hydroxide, 0.1 M, used.

4.13 Sulfuric Anhydride [13, 14, 17]

Sulfuric anhydride in the wine industry is the most commonly used preservative. There is strict legislation regarding the maximum permissible limits of sulfurous anhydride, so the control of its content in our wines has a double interest, technological as well as legislative. After adding the sulfuric anhydride to the wine it has three forms.

The first is the free sulfuric anhydride which represents the sulfite anhydride found in the form of sulfur dioxide (SO_2) as well as in the form of the inorganic sulfide compounds H_2SO_3, HSO_3^-, SO_3^-.

The second form is bound sulfite anhydride which represents sulfite anhydride bound by acetaldehyde, sugars, polyphenols, and other wine components possessing free aldehyde or ketone groups.

A third form of sulfuric anhydride in wine is total sulfuric anhydride, which is called the total amount of sulfuric acid anhydride found in wine, either free or bound.

Principle

The quantitative determination of sulfuric anhydride is based on its oxidation by iodine according to the following chemical reaction:

$$SO_2 + 2H_2O + I_2 \rightarrow H_2SO_4 + 2HJ$$

This reaction must be carried out in an acidic environment with a pH lower than 3. To calculate the free sulfuric anhydride, an iodometric determination is made directly on the wine and this is followed by a correction by repeating the iodometric determination on another sample of the same wine, since the free sulfuric anhydride of it is bound by addition of excess ethyl or propyl aldehyde.

For the calculation of the bound sulfite anhydride, the iodiometric determination is made after a double alkaline hydrolysis of the wine with NAOH 4N, whose free sulfide anhydride has already been oxidized during the preceding iodimetric determination.

Apparatus

- Conical flask of 500 mL
- Pipettes of 5, 10, 20, 30, and 50 mL
- Burette of 50 mL
- Beakers

Reagents

- Solution of NaOH 4N
- Sulfuric acid H_2SO_4 diluted ten times (180 g/L)
- Starch solution with NaCl (20%) or NaF (1‰)
- Sodium sulfate solution *N*/100
- Acetaldehyde solution 6.9 g/L
- E.D.T.A (Sodium calcium EDTA)

Procedure

In the conical flask of 500 mL we put

- 50 mL of wine
- 3 mL Sulfuric acid H_2SO_4 diluted ten times
- 30 mg E.D.T.A
- 5 mL Starch solution with NaCl (20%) or NaF (1‰)

Then titration is done with *N*/20 iodine solution until the blue color appears and remains for 15 s. We record as n the mL of iodine consumed. After that we add 8 mL solution of NaOH 4N and the mixture is kept at rest for 5 min. Then we add 10 mL of H_2SO_4 diluted ten times with vigorous stirring and repeat the titration with *N*/20 iodine solution. We record as n' the mL of iodine consumed.

After that we add to the mixture 20 mL solution of NaOH 4N and the liquid is left to stand for 5 min after stirring. When the 5 min have passed, add 200 mL of water as cold as possible and with vigorous shaking add another 30 mL H_2SO_4 diluted ten times. This is followed by another titration with iodine *N*/20 and we note up to n'' the mL of iodine consumed.

Some wine components are oxidized by iodine in an acidic environment. So when very high precision is required in our measurement, the volume of the solution consumed for the oxidation of these substances must be determined. Thus and only in this case, 50 mL of wine and 5 mL of a 6.9 g/L acetaldehyde solution or 10 g/L propanol solution are placed in a 300-mL conical bottle with a stopper. The flask is stoppered and left to rest for half an hour. After the time has passed, 5 mL of starch solution with NaCl (20%) or NaF (1‰), 3 mL of H_2SO_4 diluted ten times and we are repeating another titration with iodine *N*/20 and we note up to n''' the mL of iodine consumed.

Calculation: Expression of Results

The amount of sulfuric anhydride is expressed in mg of SO_2 and determined by approximation of 10 mg/L.

Free sulfuric anhydride is given by the following type:

$$\mathrm{SO_2L} = 32\left(n - n'''\right)\mathrm{mg\,/\,L\,SO_2}$$

Bounded sulfite anhydride is given by the following type:

$$\mathrm{SO_2C} = 32\left(n' + n''\right)\mathrm{mg\,/\,L\,SO_2}$$

Total sulfite anhydride is given by the following type:

$$\mathrm{SO_2T} = 32\left(n + n' + n'' - n'''\right)\mathrm{mg\,/\,L\,SO_2}$$

Notes

When ascorbic acid has been added to the wines we want to measure, the measurement is significantly affected. In that case with n''' we can approximately calculate the amount of ascorbic acid given that 1 mL of iodine *N*/20 oxidizes 4.4 mg of ascorbic acid.

In wines with a low sulfite anhydride content, it is recommended to use a *N*/50 iodine solution, so the number 32 in the given types becomes 12.8.

In red wines with a deep dark color, it is quite difficult to determine the color change to blue. In such cases a yellow light source from the underside helps.

4.14 Sorbic Acid [18, 19]

Sorbic acid is used in wines (mainly during bottling) to inhibit the growth of yeasts. Sorbate acts synergistically with alcohol. As in the case of sulfuric anhydride, sorbate is more active at low pH values.

Principle (Determination Using Ultraviolet Absorption Spectrophotometry)

Sorbic acid, also known as trans, trans, 2,4-hexadienoic acid, is extracted through steam distillation and its presence in wine distillate is quantified using ultraviolet absorption spectrophotometry. Substances that interfere with the measure of absorption in ultraviolet are removed by evaporation to dryness using a slightly alkaline calcium hydroxide solution. Samples containing less than 20 mg/L are verified through thin-layer chromatography, which has a sensitivity of 1 mg/L.

Apparatus

- Steam distillation apparatus
- Water bath 100 °C
- Spectrophotometer allowing absorbance measurements to be made at a wavelength of 256 nm with a quartz cell of 1 cm optical path

Reagents

- Crystalline tartaric acid
- Calcium hydroxide solution, 0.02 M
- Sorbic acid standard solution, 20 mg/L

Instruction How to Make Sorbic Acid Standard Solution, 20 mg/L
To prepare the solution, dissolve 20 mg of sorbic acid in about 2 mL of 0.1 M sodium hydroxide solution. Then, transfer it into a 1 L volumetric flask and dilute to the mark with water. As an alternative, you can dissolve 26.8 mg of potassium sorbate ($C_6H_7KO_2$) in water and make up the volume to 1000 mL with water.

Procedure
First we have to distill wine, so we place 10 mL of wine in the bubbler of the steam distillation apparatus and add about 1 g tartaric acid. Collect 250 mL of distillate.

Then we have to prepare the calibration curve, so prepare four diluted standard solutions by diluting the original solution with water, containing 0.5, 1.0, 2.5, and 5 mg of sorbic acid per liter, respectively. Use the spectrophotometer to measure their absorbance at 256 nm, with distilled water serving as the blank. Then, plot a curve to display the variation of absorbance as a function of the concentration, noting that the relationship should be linear.

For the final step of determination we place 5 mL of the distillate in an evaporating dish of 55 mm diameter, add 1 mL of calcium hydroxide solution. Evaporate to dryness on a boiling water bath. Dissolve the residue in several milliliters of distilled water, then transfer it entirely to a 20 mL volumetric flask and fill to the mark with the rinsing water. Measure the absorbance at 256 nm using a solution obtained by diluting 1 mL of calcium hydroxide solution to 20 mL with water as the blank. To determine the concentration of sorbic acid in the solution, plot the absorbance value on the calibration curve and interpolate the concentration C from it.

Calculation: Expression of Results

The sorbic acid concentration in the wine expressed in mg/L is given by the following type:

$$\mathrm{SAc} = 100 \times C$$

C = concentration of sorbic acid in the solution obtained in expressed in mg/L.

4.15 Iron

Wines contain trace amounts of iron from two different sources. One amount comes from the grapes and another amount comes from the soil that rests on the grapes and from the metal objects that the grapes, must and wine come into contact with throughout the winemaking process. More details about iron can be found in Chap. 2.

Principle

This method is based on the comparison of the color of a sample of wine, after the addition of KSCN in the presence of HCL, with a series of samples of known iron content, after being subjected to the same preparation as the sample. This coloring is due to the iron thiocyanate $Fe(SCN)_3$ produced. This is the simplest method with an accuracy of ±1 mg/L.

Apparatus

- Pipettes of 1, 10 mL
- Test tubes
- Beakers

Reagents

- Solution $FeCl_3$ containing 1 g of iron per liter

Instruction How to Make Solution $FeCl_3$ Containing 1 g of Iron per Liter

We add 1 g of pure iron or K_2SO_4, $FE_2(SO_4)_3$, $24H_2O$ or K, Fe, $(SO_4)_2$, $12H_2O$ dissolved in a mixture of 50 mL HCL and 1 mL of nitric acid. The solution is brought to a boil and after it has cooled, it is filled to 1000 mL with distilled water.

- Alcohol 95% vol.
- H_2O_2 12 w/v
- HCL Solution 0.5 N
- KSCN Solution (200 g/L)
- Diethyl ether or ether $(C_2H_5)_2O$

Procedure

Six solutions containing 3, 6, 9, 12, 15, and 18 of mg/L iron are prepared from the $FeCl_3$ solution with appropriate dilutions made by adding a hydroalcoholic solution of 12% alcohol by volume.

- *White Wines Procedure*
  Take 10 mL of each solution of different iron content and make 6 test tubes. To each one, add 1 mL KSCN Solution (200 g/L), 1 mL HCL Solution 0.5 N, and 5 drops of H_2O_2 12 w/v and mix. When we complete this process, we have created the solutions that we will compare our sample with. Then in a new test cylinder, we put 10 mL of wine and 1 mL KSCN Solution (200 g/L), 1 mL HCL Solution 0.5 N and 5 drops of H_2O_2 12 w/v. After making sure that all the test tubes are homogenized we compare the sample with the test tubes of known content.
- *Red Wines Procedure*
  Take 10 mL of each solution of different iron content and make six test tubes. To each one, add 10 mL Diethyl ether or ether $(C_2H_5)_2O$, 1 mL KSCN Solution (200 g/L), 1 mL HCL Solution 0.5 N, and 5 drops of H_2O_2 12 w/v and mix. When we complete this process, we have created the solutions that we will compare our sample with. Then in a new test cylinder we put 10 mL of wine, 10 mL Diethyl ether or ether $(C_2H_5)_2O$, 1 mL KSCN Solution (200 g/L), 1 mL HCL Solution 0.5 N, and 5 drops of H_2O_2 12 w/v. After making sure that all the test tubes are homogenized we compare the sample with the test tubes of known content.

Calculation: Expression of Results

We identify which controls match the color of the sample and then reduce to the liter. We must also take into account any dilutions of our wine.

> **Note**
> In the extreme case that we have a stronger coloration of our sample than the coloration of the tube containing 19 mg/L of iron, repeat the process by diluting the wine sample twice with distilled water. In the end, however, to reduce to liter we have to multiply by 2.

4.16 Potassium

Potassium is contained in wines in fairly large quantities compared to the rest of the cations. The concentration in general usually ranges from 0.1 to 1.8 g/L.

Principle (Zygometrical Determination)

The principle of this method is based on the precipitation of the potassium of the wine by the addition of sodium tetraphenylboride in an acidic environment. Additionally, the value of the method can be confirmed by comparative volumetric determination. The accuracy of the method is ±0.02 g/L.

Apparatus

- Pipettes of 0.5, 2, 5, 10, 20, and 25 mL
- Volumetric flask of 25 mL
- Platinum beaker
- Porcelain filter (pore diameter no. 4)
- Electric furnace temperature 525 ± 25 °C
- Furnace 110 °C
- Water bath

Reagents

- Sodium tetraphenylborate solution 3%
- Saturated aqueous solution of bromocresol green
- Sulfuric acid 10%
- Potassium-free NaOH solution 1N

Procedure

In the platinum beaker, we put 25 mL of wine which are placed in the water bath and evaporated. The residue is charred in a furnace at 525 ± 25 °C. The ash formed after its carbonization is taken up with 20 mL of distilled water and acidified with 0.5-mL sulfuric acid 10%. The whole is transferred to a 25-mL volumetric flask in which the water for rinsing the capsule is also transferred. It is filled up to 25 mL which is accompanied by good stirring and filtered.

Take 10 mL of filtrate and add 1 drop of saturated aqueous solution of bromocresol green and such a quantity of potassium-free NaOH solution 1N, that the solution acquires the green color of bromocresol (pH 4–5). Then heat in a water bath to 50 °C and add 5 mL sodium tetraphenylborate solution 3%. The addition should be done drop by drop and with constant stirring.

It is then allowed to cool and the resulting white precipitate is filtered in porcelain filter (pore diameter no. 4) and then wash the platinum beaker and the precipitate with water containing 2 mL of sodium tetraphenylborate solution 3% solution per 100 mL of water. The porcelain filter is drained well, dried at 110 °C and accurately weighed.

The porcelain filter is then rinsed three times with acetone, dried and accurately weighed again. The difference between the two weights is noted as *P* and is equal to the weight of potassium tetraphenylboride.

Calculation: Expression of Results

The amount of potassium is given by the following types:

$$\mathrm{Kw} = 279.70 \times \mathbf{P}\,\mathrm{meq/L}$$

$$\mathrm{Kw} = 10.91 \times \mathbf{P}\,\mathrm{g/L\ Potassium}$$

$$\mathrm{Kw} = 52.64 \times \mathbf{P}\,\mathrm{g/L\ Potassium\ tartrate}$$

4.17 Phenolic Compounds

It is difficult objectively to express with a number the set of phenolic substances. For this reason, there are certain indicators that express in a conventional and not absolute way the total content of phenols in the wine.

4.17.1 Potassium Permanganate Index

Principle

This method is based on the oxidation of the phenolic compounds of potassium permanganate at ambient temperature. The end of the reaction is marked by the color change of the carmin-indigo indicator from blue to yellow. This indicator oxidizes before certain wine components, but after phenolic compounds. Its color change coincides with the end of the oxidation of phenolic compounds, so as to avoid further consumption of potassium permanganate to oxidize other indifferent components of the wine.

Apparatus

- Volumetric flask of 500 mL
- Burette of 50 mL
- Beakers of 250 mL
- Pipettes of 1, 25, 50 mL

Reagents

- Carmin-indigo solution 3%
- Sulfuric acid solution 30%
- Liqueur carmin-indigo

Instruction How to Make Liqueur Carmin-Indigo

In a volumetric flask of 500 mL, we add 25 mL of carmin indigo solution 3%, 25 mL of sulfuric acid 25%, and distilled water until 500 mL.

- Potassium permanganate solution ($KMnO_4$) *N*/100 (0.316 g/L)
- Tartaric acid solution 5 g/L which is 1/3 neutralized and contains 10% alcohol
- Wine sample

Procedure

In a beaker, we add 50 mL of Liqueur carmin-indigo and 1 mL of wine sample. Then with the burette, we add potassium permanganate solution ($KMnO_4$) *N*/100 until the blue color became pale yellow. We note the volume of $KMnO_4$ as *n* mL.

A part of $KMnO_4$ *N*/100 is consumed by tartaric acid, therefore we repeat the previous determination with the difference that instead of adding 1 mL of wine, we add 1 mL of a 5 g/L solution of tartaric acid which has been neutralized by 1/3 and which contains 10% alcohol. We do the titration and note with n' the consumption of $KMnO_4$ *N*/100.

Calculation: Expression of Results

The content of the wine in phenolic compounds is given by the potassium permanganate index (Ip) and is expressed in meq/L by the following type:

$$\text{Ip} = 10 \times (n - n')\,\text{meq / L}$$

The number resulting from the application of the formula when it is less than 30 meq/L, we consider that the wine we examined is soft. When the result is between 30 and 50 meq/L, we have wines full of body and when the reading is more than 50 meq/L we have astringent wines.

4.17.2 Folin–Ciocalteu Indicator

Principle

The method is based on the oxidation of phenolic compounds by the Folin–Ciocalteu reagent. During the oxidation of phenolic compounds, the acids are reduced to a mixture of blue oxides, namely tungsten and molybdenum. The blue coloration shows the absorption maximum at 600–760 nm. The reaction takes place in an alkaline environment and the intensity of the blue color is proportional to the amount of phenols.

The disadvantages of the method are that the color formed is not stable and evolves over time, specifically after the first 30 min the blue turns into a deep blue. This means that for our results to be comparable the absorbance measurements must be done at a constant time for all samples.

Apparatus

- Spectrophotometer
- Volumetric flask of 100 mL
- Pipettes of 1, 5, 10 mL

Reagents

- Folin–Ciocalteu reagent
- Anhydrous solution Na_2CO_3 20%
- Wine Sample

Procedure

In a volumetric flask of 100 mL, we add 5 mL of Folin–Ciocalteu reagent, 10 mL of Anhydrous solution Na_2CO_3 20%, and 1 mL of wine. In white and rosé wines, the samples are used without dilution. Flowing red wines are diluted at a ratio of 1/10 while red pressure wines are diluted at a ratio of 1/20. The volumetric flask is filled with distilled water up to the 100 mL mark.

Mix and leave to rest for 30 min. Just after the end of 30 min and certainly before 45 min have passed, the absorbance of the sample is measured in the spectrophotometer at 700 nm in a 1-cm cell.

Calculation: Expression of Results

Folin–Ciocalteu indicator is given the following type:

$$I_{CF} = (\text{Absorption} \times \text{Dilution}) \times 20$$

The number resulting from the application of the formula when it is less than 30 meq/L, we consider that the wine we examined is soft. When the result is between 30 and 50 meq/L, we have wines full of body, and when the reading is more than 50 meq/L, we have astringent wines.

4.17.3 Index of Total Polyphenols (D.O 280 nm)

Principle

The method is based on the principle that the benzene rings of the phenolic compounds of the wine show strong absorption in ultraviolet light, the maximum of which is observed at 280 nm.

Apparatus

- Spectrophotometer U.V
- Volumetric flask of 100 mL
- Pipette of 1 mL

Reagents

- Wine sample

Procedure

Dilute the red wine sample with distilled water 100 times and homogenize. White wines are diluted ten times with distilled water. We then measure the optical density in a 1-cm cell at 280 nm and we note D.O.

Calculation: Expression of Results

The index of total polyphenols is obtained from the following formula:

$$D_{280} = \text{D.O} \times \text{Dilution}$$

The resulting value is between 6 and 120. Values between 60 and 80 characterize a rich full-bodied wine with a deep red color. Index D_{280} and Folin–Ciocalteu index are linked by the following relation:

$$D_{280} / I_{\text{FC}} = 1.2 - 1.3$$

4.17.4 Anthocyanin Concentration with the Method of pH Variation

Principle

The method is based on the property that anthocyanins have to change their color depending on pH.

Apparatus

- Spectrophotometer U.V
- Volumetric flask of 100 mL
- Pipette of 1 and 10 mL
- Burettes

Reagents

- Hydrochloric Acid in Ethanol Solution Hydrochloric Acid 0.1%
- 3.5 pH buffer
- HCl Solution 2%

Procedure

In a burette (A), we add 11 mL of wine sample, 1 mL of Hydrochloric Acid in Ethanol Solution Hydrochloric Acid 0.1%, and 10 mL HCl solution 2%. In a second burette (B), we add 11 mL of wine sample, 1 mL of Hydrochloric Acid in Ethanol Solution Hydrochloric Acid 0.1%, and 10 mL of 3.5 pH buffer. Then we measure the optical density at 520 nm, for solution A and then again for solution B. We note the optical density in solution A as n_1 and the optical density in solution B as n_2. Finally we calculate their difference and write it as $\Delta_{n_1 - n_2}$.

Calculation: Expression of Results

To calculate the concentration of anthocyanins we apply the formula. The resulting number can be compared with solutions of known origin in anthocyanins.

$$C = \Delta_{n_1 - n_2} \times 388\text{mg} / \text{L}$$

To be more specific, anthocyanins in wine are divided into free anthocyanins and those bound to tannins. Of those that are bound to tannins, some are discolored by sulfur dioxide and some are not discolored. It should be noted that with this method the concentration of free anthocyanins and anthocyanins decolorized by sulfur dioxide is calculated.

4.18 Color Evaluation of Wines

4.18.1 Red Wines (Sudraud Method)

The color of the wines is due to the phenolic compounds that come from the solid parts of the grape. These components through extraction pass into the must and then into the wine. These components, in addition to the color, also affect the organoleptic characteristics of the wines. From the above we understand that the existence of objective indicators in the evaluation of the color of wines is very useful for many reasons. To assess the color of red or wines we use the **intensity** (*I*) and the **hue** (*H*). These two indicators can be measured very easily using a spectrophotometer. To be more specific, intensity is defined as the sum of the optical densities measured at the wavelengths of 420 and 520 nm. Hue is defined as the quotient of the optical densities at the wavelengths of 420 and 520 nm.

Principle

The method is based on the fact that color is the result of the selected absorption of elementary rays that make up the solar spectrum. For white wines and their shades, it has been established that the maximum absorption is observed in ultraviolet light, specifically at a wavelength of 270–280 nm. For shades related to oxidized white wines, the absorption measurement is indicated at 420 nm. Red wines, due to the large color palette, it is necessary to measure at three wavelengths, specifically at 620 nm where we have maximum absorption of the purple color, at 520 where we have maximum absorption for the vibrant red color and at 420 nm where the absorption of yellow color shows some increase.

Apparatus

- Spectrophotometer U.V
- 10 mm quartz cuvette

Reagents

- Wine s.ple

Procedure

The process of measuring optical densities is very simple. Specifically for red wines, the measurement is made in relation to distilled water in the spectrophotometer in quartz cells with an optical path of 10 mm at wavelengths of 420, 520 nm.

Calculation: Expression of Results

$$I = D_{420} + D_{520}$$
$$H = D_{420} / D_{520}$$

4.18.2 White Wines

The assessment of the color of white wines is more complex than that of red. This is because in white wines the spectrum does not have a defined maximum in the visible spectrum. Absorption is continuous from 500 to 280 nm with a maximum in the UV. From the above we understand that it is difficult to rely on the spectrum for the interpretation of the color of white wines. The color yellow has a characteristic absorption at 420 nm and this value alone gives an approximate indication of the color. White wines therefore have a peculiarity regarding the determination of color. But there are two useful indicators that by determining them we avoid future problems related to color development. The first is the browning index, with its determination we can prevent coloring problems due to oxidation. The second index is the redness index which shows us how prone a white wine is to acquire a characteristic reddish color in the future.

4.18.2.1 Browning Index

Principle

The analysis is based on the effect that oxygen has on white wines, causing them to acquire brown to brown shades after being exposed to it.

Apparatus

- Spectrophotometer U.V
- 10 mm quartz cuvette
- Incubation chamber

Reagents

- White wine sample

Procedure

Twenty milliliter of white wine is filtered through a 0.45 μ membrane filter and aseptically placed in a screw-capped test tube. It is important that the test tube is sterilized before introducing the wine in order to avoid any contamination that will lead us to wrong conclusions. Then we measure the optical density of the sample at 420 nm using the spectrophotometer in a 10-mm quartz cell. The optical density is denoted as D_1. Then we pass pure oxygen to the sterilized test tube for a duration of 10 min. After 10 min, screw the cap of the test tube and put it in an incubation chamber for 9 days at 55 °C. When the 9 days have passed, we repeat the measurement of the optical density and note it as D_2.

Calculation: Expression of Results

$$|D_2 - D_1| \times 1000$$

We multiply the result obtained from the difference in measurements by 1000. The higher the result, the more elixir the wine is. When we have results greater than 70, the wine must be treated with special oenological material that will be analyzed in the next chapter to avoid this problem.

4.18.2.2 Redness Index

Principle

The analysis is based on the effect of reaction of phenolic compounds, such as proanthocyanidins, with oxygen and metals like iron or copper.

Apparatus

- Spectrophotometer U.V
- 10-mm quartz cuvette

Reagents

- White wine sample

Procedure

Take a sample of the white wine and divide it into two parts: a control and a test sample. Add 0.3% hydrogen peroxide to the test sample to induce oxidation. Let the samples rest for 24 h. Measure the optical density at 500 nm for both samples D_1 and D_2 using a spectrophotometer.

Calculation: Expression of Results

The difference between the control D_1 and the test sample D_2 indicates the wine's sensitivity to pinking. A higher difference suggests a greater susceptibility to pinking.

$$|D_2 - D_1|$$

4.19 Composition and Methods of Preparation of Special Reagents [20, 21]

In this paragraph, we will show how some reagents that we use in analytical chemistry are prepared and are named after the first scientist who prepared them. It is very common in the analysis protocols available in the literature or on the Internet to refer to these solutions with names such as (reagent), without mentioning the methods of preparation of the reagents. From personal experience, this fact creates a confusion and an insecurity in the already difficult task for the person charged with the difficult process of laboratory analyses. For this reason, we decided to gather in this chapter those solutions related to the science of oenology and to analyze step by step how they are prepared.

- *Fehling's Solution*
  Fehling's solution is created by mixing two distinct solutions: Fehling's A, a deep blue aqueous solution containing copper(II) sulfate, and Fehling's B, a colorless solution of potassium sodium tartrate (Rochelle salt) in water, rendered strongly alkaline with sodium hydroxide.
  To create Fehling's A, we add 69.28 g of crystal $CuSO_4$ and we dissolve in 300 mL of distilled water and fill up with distilled water to 1000 mL. To create Fehling's B, we add 346 g of sodium potassium tartrate and 120 g of NaOH in 500 mL of distilled water and fill up with distilled water to 1000 mL.
- *Folin–Marenzi Solution*
  The Folin–Marenzi reagent, also referred to as the Folin–Ciocalteu reagent, consists of phosphomolybdate and phosphotungstate. It is mainly utilized for the colorimetric in vitro assay of phenolic and polyphenolic antioxidants. This reagent plays a crucial role in the Lowry method for protein concentration determination and is employed in the quantification of total phenolic content.
  To create Folin–Ciocalteu reagent, we add 100 g of sodium tungstate and 25 g of sodium molybdate are dissolved in 700 mL of water, 50 mL of 85% phosphoric acid, and 100 mL of concentrated HCl with specific gravity 1.19 are added to the solution. Place the solution in a boiling flask with a vertical cooler and boil for 8 h continuously. If the solution turns green add a few drops of bromine water [Bromine water is an oxidizing, intense brown mixture containing diatomic bromine (Br_2) dissolved in water (H_2O)] until it turns yellow, with a little boiling the excess bromide evaporates. If a precipitate forms, the solution should be filtered. Finally, it is filled with distilled water up to 1000 mL.
- *Griess–Ilosvay Reagent*
  The Griess–Ilosvay reagent is a chemical solution utilized for the detection of nitrites, commonly indicative of microbial activity. In microbiology, this reagent is especially valuable for pinpointing nitrate-reducing microorganisms. The presence of nitrites in a sample triggers a reaction with the reagent, resulting in a red or pink hue, which simplifies the visual confirmation of their presence.
  To create Griess–Ilosvay reagent, 5 g of sulfanilic acid and 3 g of α-naphthylamine are dissolved in hot water. Then add 10 cm^3 of concentrated acetic acid (CH_3COOH) and fill with distilled water up to 1000 mL.
- *Mayer Reagent*
  Mayer's reagent, an alkaloidal precipitating agent, is utilized for detecting alkaloids in natural substances. To make Mayer solution, we must first make two different solutions and then mix them. For the first solution (A), we add 27 g of $HgCl_2$ to 1000 mL of warm distilled water. For the second solution (B), add

100 g of Potassium Iodide to 500 mL of hot distilled water. For the final solution, mix solutions (A) and (B) and add 25 mL of concentrated acetic acid (CH_3COOH).

- *Nessler Reagent*
  We use Nessler reagent to detect ammonium salts. This pale solution becomes deeper yellow in the presence of ammonia NH_3. To prepare it, follow the procedure below. First dissolve 50 g of potassium iodide KI in 100 mL of distilled water. Then dissolve 25 g of mercuric chloride $HgCl_2$ in 400 mL of distilled water. This may take some time as mercuric chloride is not very soluble in water. Slowly add the mercuric chloride solution to the potassium iodide solution while stirring continuously. A precipitate of mercuric iodide HgI_2 will form initially but will dissolve as more mercuric chloride is added, forming potassium tetraiodomercurate (II) K_2HgI_4. Dissolve 150 g of potassium hydroxide KOH in 500 mL of distilled water. Once fully dissolved, add this solution to the mixture of potassium iodide and mercuric chloride. Dilute the final mixture to 1 L with distilled water. Ensure the solution is well mixed. Store the Nessler's reagent in a dark bottle to protect it from light, which can cause decomposition.
- *Nylander Reagent*
  Nylander's reagent is utilized for detecting the presence of reducing sugars. Under alkaline conditions, glucose or fructose reduces bismuth oxynitrate to metallic bismuth. When Nylander's reagent, composed of bismuth nitrate, potassium sodium tartrate, and potassium hydroxide, is introduced to a solution containing reducing sugars, a black precipitate of bismuth is produced. o prepare it, follow the procedure below. First dissolve 2 g of bismuth nitrate BiN_3O_9 in 10 mL of distilled water. This may require gentle heating to fully dissolve. Then dissolve 4 g of potassium sodium tartrate $C_4H_4O_6KNa \bullet 4H_2O$ in 10 mL of distilled water. Slowly add the potassium sodium tartrate solution to the bismuth nitrate solution while stirring continuously. Dissolve 8 g of potassium hydroxide KOH in 20 mL of distilled water. Once fully dissolved, add this solution to the mixture of bismuth nitrate and potassium sodium tartrate. Dilute the final mixture to 100 mL with distilled water. Ensure the solution is well mixed. Store Nylander's reagent in a dark bottle to protect it from light, which can cause decomposition.

4.20 Calculation of Calories in Wine

To calculate the calories contained in a wine in kcal, we need to know the alcohol content of the wine, the residual sugars included in the wine and the SFE. Sugar-free extract and the closely related "total dry extract" (TDE) are traditional measures used in winemaking. TDE is defined as "all matter that is non-volatile under specified physical conditions" [22]. To illustrate, if a wine sample were evaporated under controlled conditions, the remaining mass, typically a slightly sticky solid, would be the TDE. This residue primarily consists of organic acids, glycerol, and sugars [23], but may also include substantial amounts of other components like tannins and proteins, varying with the type of wine. The sugar-free extract (SFE) is calculated by subtracting the sugar mass from the TDE.

Sugar-free extract index given by the formula below:

$$\text{Sugar} - \text{free extract} = \text{total dry extract} - \text{total sugars}$$

So since we know the above, we can apply it to the formula for calculating calories in Kcal per 100 mL given below and find the calorific value of the wine in Kcal:

$$\text{Kcal}(\text{calories}) = 6.9 \times 0.794 \times \%\text{alcohol by volume} + 4.0 \times (\text{g sugar}/100\,\text{mL}) + 2.4 \times (\text{g sugar} - \text{free extract}/100\,\text{mL})$$

If we want to express it in Kj we simply multiply the result we have in Kcal by 4.184 and express it in Kj.

References

1. International Organisation of Vine and Wine (OIV) (2021) Compendium of international methods of wine and must analysis, vol 1. OIV, OIV Compendium
2. Semichon L, Flanzy M (1933) Ann Fals Fraudes 26:404
3. Peynaud E (1951) Bull Soc Chim Biol 18:911. [Ref. Z. Lebensmit. Forsch. 1953; 97:142]
4. Pato M (1944) Bull O.I.V., 17, no, 161, 59, no, 162, 64
5. Poux C (1949) Ann Fals Fraudes 42:439
6. Jaulmes P, Brun MS, Vassal MM (1961) Trav Soc, Pharm, Montpellier 21:46–51
7. Jaulmes P, Brun MS, Cabanis JC (1969) Bull OIV, nos 462–463, 932
8. Reinhard C, Koeding G (1989) Zur Bestimmung der Apfelsäure in Fruchtsäften. Flüssiges Obst 45:S, 373 ff
9. Afnor standard, 76–107, Arnor, Tour Europe, Paris. Prom T., F.V., O.I.V. (1984), no 788
10. Pien J, Meinrath H (1938) Ann Fals Fraudes 30:282
11. Dupaigne P (1947) Bull Inst Jus Fruits. No 4
12. Tavernier J, Jacquin P (1947) Ind Agric Alim 64:379
13. Jaulmes P, Hamelle Mlle G (1956) Mise au point de chimie analytique pure et appliquée, et d'analyse bromatologique, par J.A. Gautier, Paris, 4e série
14. Jaulmes P, Hamelle Mlle G (1963) Trav Soc Pharm Montpellier:243
15. Hamelle Mlle G (1965) Extrait sec des vins et des moûts de raisin, Thèse Doct. Pharm. Montpellier
16. Jaulmes P (1951) Analyse des vins, Librairie Poulain, Montpellier, éd., 107
17. Soufleros E (2012) Oenology, science and technology, Thessaloniki
18. Jaulmes P, Mestres R, Mandrou B (1961) Ann Fals Exp Chim, n° spécial, réunion de Marseille:111–116
19. Chretien D, Perez L, Sudraud P (1980) F.V., O.I.V., no 720
20. Treadwell: Chimie analytique I. Analyse qualitative
21. Chapler: Formulaire du Laboratoire
22. OIV (2015) Compendium of international methods of wine and must analysis, vol 1. OIV, Paris
23. Boulton RB (1996) Principles and practice of winemaking. Chapman & Hall, New York, pp 138–139

Oenological Additives

5

5.1 Yeasts

5.1.1 General Information

Yeasts have perhaps the most important role in the winemaking process. These are responsible for converting the grape must into wine and more specifically the sugars contained in the must into ethanol. Yeasts range in size from 5 to 8 μ and are a group of unicellular organisms and are referred to by their Latin genus name and species name. The term yeasts refer to a group of unicellular organisms that are neither fully defined nor homogeneous. However, they share many characteristics, the most important of which is that they cause various fermentations. By fermentation we mean the breaking down of complex organic components into other simpler components.

Yeasts belong to eukaryotic cells. Cells that have a fully formed nucleus are called eukaryotic, as opposed to cells that do not have a formed nucleus, called prokaryotic cells. Basic parts of a eukaryotic cell are (a) plasma membrane, (b) cytoplasm and cell nucleus. Examining the structure of these cells, it is observed that externally they are surrounded by a membrane and internally the nucleus is separated from the rest of the cell again by a membrane (nuclear membrane). Between the nucleus and the outer membrane are organelles, which are responsible for the various functions of the cell, such as mitochondria, lysosome, and ribosome.

It is known that all organisms whether unicellular or multicellular are made up of cells. Cells grow in an aqueous environment, and this implies a tendency for water to flow into its interior. This phenomenon is called osmosis. Through this phenomenon it tends to balance the concentration of solutes outside and inside the cell. So, to compensate for this flow, cells have developed the cell wall. This membrane, called the cell wall, constitutes 15–25% of the dry weight of the cell, is highly elastic and consists mainly of polysaccharides (β-glucan, mannoprotein, and chitin).

G. Z. Kyzas, F. Papageorgiou, *Oenology in Practice*,
https://doi.org/10.1007/978-3-031-85531-3_5

Beneath the cell wall is the plasma membrane. It controls the exchange of metabolites with the external environment. The accepted model for its structure and behavior of the plasma membrane is that of the fluid mosaic. This model states that the cell membrane consists of a fluid lipid bilayer with proteins embedded in it. It emphasizes the fluid nature of the structure in which lipids and proteins can move laterally within the membrane as well as its mosaic nature as it contains a variety of proteins and lipids, as well as some carbohydrates, that are unevenly distributed. This model recognizes that the membrane is selectively permeable, allowing some molecules to pass while restricting others. The selective permeability characteristic of the cell membrane is crucial to its role in cellular homeostasis, regulating the entry and exit of substances between the cell and its environment. The plasma membrane of *Saccharomyces cerevisiae* consists of 40% lipids, 50% proteins, and a small percentage of glucans and mannans.

The inner part of the cell—outside the nucleus—surrounded by the cell membrane is called the cytoplasm. It is characterized as a viscous material (like jelly) which holds the rest of the organelles, protects them, and replenishes the empty space of the cell. New studies have shown that the cytoplasm of eukaryotic cells is traversed by a polymorphic network of fibrils that make up the cytoskeleton. Thanks to the cytoskeleton, they are mechanically supported and can maintain as well as change their shape.

The classification has changed many times to date. Of the 4000 types of yeasts that have been classified, only 15 of them are involved in the winemaking process. In the beginning, the classification of yeasts was limited to the difference in shape, dividing the yeasts into ellipsoid (ellipsudeus) and apiculata (apiculata). Later the ability to ferment various sugars came into use. At that time, a distinction was made between the species *Saccharomyces cerevisiae* (previously included in the ellipsoid) and *Saccharomyces oviformis*, which was considered more resistant to alcohol. In addition, the former fermented galactose while the latter did not. The *Saccharomyces bayanus* species had the same physiological characteristics as *Saccharomyces oviformis* regarding the fermentation of sugars, but had different morphological characteristics, slower fermentation, and different fermentation behavior. At the same time, the species *Saccharomyces uravum* had been identified, unlike the aforementioned, it could contain melibiosis. So a new classification followed in 1970 and various modifications were made as a result of which some species were moved to different genera and some other species were considered synonyms and joined the same category. At the end of the same decade, the API 20 classification system began to be used, followed by API 50, and later another classification based on the percentage of the amino acids guanine and cytosine in the DNA of yeasts brought new changes. In this new classification method, only 7 species were retained, while 17 species were considered strains of *S. cerevisiae* and their differentiation was now based on the ability to ferment sugars. In 1990, a new classification was based on DNA recombination tests and crossover experiments. Then the species became 10 and were divided into three groups. With this classification, a confusion was then created as according to biologists *Saccharomyces cerevisiae* and *Saccharomyces bayanus* are considered different species, while in oenology, they still consider

Saccharomyces bayanus to be a strain of the cerevisiae species. Also, most strains used in winemaking and in a previous classification were strains of the species *S. uravum* now belong to the species *Saccharomyces bayanus*. Research in 2000 has added three more species, namely *Saccharomyces cariocanus*, *Saccharomyces kurdianzenii*, and *Saccharomyces mikatae*. In summary, we must admit that the development of science does not aim to create confusion, on the contrary, it aims at a more precise classification, but the confusion is created as scientists have to revise data that they may have embraced throughout their professional career.

In winemaking today, if we decide to intervene with the addition of commercial yeasts and not let native yeasts carry out the alcoholic fermentation, we can choose from a wide variety of yeasts. The choice of yeasts depends on the wine we have decided to create. Each winemaking yeast company develops and improves yeasts with the help of genetic engineering and markets them under a different brand name. From our side, once we know our raw material, we must choose the most suitable one to create the wine we desire. Below are the yeast strains one may encounter in winemaking (desired or unwanted) as well as the most common commercial yeast options and their main characteristics [1].

5.1.2 Yeast Strains Related to Winemaking

5.1.2.1 *Acremonium* sp.

Infection in wine from *Acremonium* sp. can lead to the formation of TCA (2,4,6-trichloroanisole) and TBA (2,4,6-tribromoanisole). These compounds are implicated in the "corked" character of some wines.

5.1.2.2 *Alternaria alternate*

It is the flora of the vineyard and the winery. Possibly present with other molds on wet/damaged fruit. There is a possibility that it is contributing to the rot. It does not grow during the wine fermentation process but is likely to cause mold in wines made from infected grapes. This particular strain poses more of a health risk to workers as it has been associated with asthma symptoms in US homes and can cause skin rashes.

5.1.2.3 *Aspergillus* sp.

Aspergillus rot is caused by different members of the genus *Aspergillus* which are widespread throughout the world. In grapes, especially *A. alliaceus*, *A. carbonarius*, *A. niger*, and *A. ochraceus* occur. The genus *Aspergillus* includes more than 200 species including species with many subspecies. Some of these subspecies produce mycotoxins that are carcinogenic and make the fungi important to exclude from winemaking [2]. *Aspergillus* is usually a large part of what is referred to as mildew or mold. Some people have developed severe allergies to *Aspergillus*, so it is important to try to remove any visible colonies and prevent mold regrowth to protect workers and consumers in a winery. *Aspergillus* that uses the must for substrate and must be effectively managed because the final wine can be significantly altered by the sugar metabolism of the budding wine. This glucose oxidation can have a profound esthetic effect as well as reduce the alcohol content of the final wine by sequestering the substrate that Saccharomyces needs to produce alcohol. In the vineyard, appropriate measures should be taken, and the proliferation of organisms should be reduced. Measures such as removing excess leaves and reducing the number of crops so that clumps are not in close contact will allow improved air flow while reducing moisture content. Pulling the leaves will also allow sunlight to reach the berries and improve spray penetration. The winery can also be a source of mold. Winery equipment must not have porous surfaces. Everything should be cleaned frequently with substances known to be safe for the wine and the consumer, but still take advantage of mold sensitivity to inhibit mold growth [3].

5.1.2.4 *Brettanomyces*

This particular strain imparts strong flavors and aromas to the wine that are often described as horse blanket, barn, aid, and excrement. Except in special cases of wines, their development is undesirable because it forms a significant amount of ethyl acetate and acetamide.

5.1.2.5 *Candida intermedia*

C. intermedia is not commonly found in the microflora of grapes or winery environments; it is more often associated with humans. While it has the potential to cause spoilage, it is not a common spoilage yeast in wine production.

5.1.2.6 *Candida krusei*

Can co-ferment a wine, especially at low temperature. Presence in wine contributes to slime formation and ropiness.

5.1.2.7 *Candida parapsilosis*

Can be part of natural grape flora. Can ferment to a limited degree.

5.1.2.8 *Candida stellate*

Candida stellata is part of the natural grape and fermentation flora. Generally, they are most abundant at the beginning of fermentation, but as the process advances, their numbers diminish, yielding to *Saccharomyces* species. However, *C. stellata* and other *Candida* species have been found present much later in the fermentation process than other wild or native yeast and have been shown to completely ferment a wine. *C. stellata* may manifest as a white, cheese-like film on the surface of wine, yet its impact is largely esthetic, as it does not appear to cause significant negative sensory or stability issues. Indeed, certain studies have shown that *C. stellata* can improve the sensory profile of wines, imparting flavors such as honey, apricot, and sauerkraut. *C. stellata* is not only more prevalent at the beginning of fermentation but also thrives in wines that ferment at lower temperatures. The yeast population grows in wines undergoing cold soak and those kept at low temperatures for solid settling.

5.1.2.9 *Candida vini*
Spoilage yeast, which makes wines sour and/or cloudy.

5.1.2.10 *Candida zemplinina*
Isolation of organisms from overripe and botrytized grapes is possible because of their tolerance to high sugar levels and low temperatures, conditions that are lethal for many other organisms. This yeast is capable of surviving in both must and wine, with populations peaking during the initial stages of fermentation, yet it can complete the alcoholic fermentation process. It is possible for this organism to exist during storage, as other *Candida* species can be present in barrels, producing a surface film.

5.1.2.11 *Cryptococcus laurentii*
Can occur as native fungi on grape skin and also found in cellars and in corks; Rarely present in finished wine in significant quantities. In corks, it may be responsible for some odor/contamination problems; recent research suggests that certain methods may serve as effective biocontrol in vineyard management, particularly against Botrytis cinerea, due to their ability to degrade laccase.

5.1.2.12 *Debaromyces hansenii*
Normal grape/fermentation flora produces high levels of β-glucosidases, which creates more monoterpenol in the wine, affecting sensory results. Sensory effects include higher terpenol flavors, especially linalool.

5.1.2.13 *Elsinoe ampelina*
In vineyards, *E. ampelina* can diminish the lifespan of vines, sap their vigor, and lower both the yield and quality of the fruit. Berries infected by this pathogen can adversely affect the must's quality, thereby impacting the wine's quality as well. The infected berry will crack and then be susceptible to secondary infections which would also have an effect on the juice/wine quality [4].

5.1.2.14 *Guignardia bidwellii*
Cluster infections impact both the yield and the quality of berries, rendering the affected grapes unsuitable for wine production. It is advisable to refrain from using grapes infected with black rot in winemaking [5].

5.1.2.15 *Hanseniaspora uvarum*
This particular strain as well as the imperfect form of Kloeckera apiculata is found in the form of natural flora on the grape. Its growth and proliferation is inhibited at amounts of sulfuric anhydride of 100 mg/L and above. It has a low fermentation efficiency of up to 4% vol. It dominates at the beginning of fermentation since it constitutes 90% of the yeast population in the grape, but due to its low fermentation efficiency and its sensitivity to anhydride sulfite, other yeast strains with greater fermentation efficiency and resistance to alcohol and anhydride sulfite quickly disappear as well the strain *Saccharomyces cerevisiae* or *Saccharomyces bayanus.*

Kloeckera apiculata strain creates increased volatile acidity as well as ethyl acetate. More generally, it is an undesirable strain and we must take the necessary measures to avoid its development.

5.1.2.16 *Hansenula anomala*

Hansenula anomala is found in the normal grape flora and is active in early fermentation. Appears in grapes infected with Botrytis. It has the ability to form chains by joining the cells and a thin membrane on the surface of the wines. It can spoil wine by excessive production of acetic acid and ethyl acetate, sediment formation and acid metabolism—increase in pH—. This particular strain is considered highly undesirable. It has the property of producing an effective killing agent against Dekkera/*Brettanomyces* spp. [6].

5.1.2.17 *Penicillium expansum*

Disease agent in vines that may impart an off-flavor to the wine. Reports indicate that even small quantities of infected berries can lead to a moldy flavor in wine.

5.1.2.18 *Penicillium* sp.

It is an undesirable strain in viticulture and winemaking. Huge negative effect on wine feel as well as possible increase in titratable acidity. *Penicillium* spores contain mycotoxins (patulin, citrin, roquefortin, ochratoxin A—"OTA") responsible for causing a variety of allergy symptoms and diseases. The International Agency for Research on Cancer has classified OTA as a probable human carcinogen. Fungi have the ability to alter the chemical makeup of berries, influencing yeast development during the alcoholic fermentation process, which can result in variations in the wine's color and taste [7].

5.1.2.19 *Pichia etchellsii*

Some species of Pichia, in the general definition before splitting, can disrupt the fermentation process in the production of alcohol. In winemaking, some Pichia species can create potential defects in wines. Most are found in decaying plants. Some live in close symbiosis with insects, which live on decaying plants. This particular strain has a low fermentative efficiency and creates increased volatile acidity and ethyl acetate. It also has the ability to form pseudomycelia and grows on the surface of wines creating a thin film.

5.1.2.20 *Pichia guilliermondii*

Wine stored in barrels can be affected by certain organisms such as this one, leading to organoleptic defects. The lesion may appear as a white foam on the surface. It can also cause harmful concentrations of ethyl acetate.

5.1.2.21 *Rhizopus stolonifer*

This Fungus spores can be present in vineyards, particularly in warm or moist environments, residing in the soil or on the berries themselves. Should the berries sustain damage, these spores are capable of penetrating the fruit's flesh and consuming its sugars. Using moldy grapes to make wine will result in a high production of ethyl acetate, which is a volatile compound with an unpleasant aroma, undesirable in wine.

5.1.2.22 *Saccharomyces baili*

This specific strain is rarely found in the grape. It is usually found in sweet wines and is responsible for fermentations in wines that have not undergone the proper care required in this type of wine [1].

5.1.2.23 *Saccharomyces bayanus*

Saccharomyces bayanus is present in small quantity in the grape. It is quite resistant to ethanol and can ferment up to 18% vol. This is the yeast that is responsible for 50% of fermentations and is resistant to sulfuric anhydride.

5.1.2.24 *Saccharomyces cerevisiae*

Saccharomyces cerevisiae is the most important yeast species involved in wine fermentation. Traditionally, this species has been used as a starter culture to carry out alcoholic fermentation due to its optimal fermentation properties. *Saccharomyces cerevisiae* makes up 80% of the fermenting must population, it multiplies quickly and has a fermentation efficiency of up to 14% vol, it cannot ferment galactose and slightly increases volatile acidity [8]. *S. cerevisiae* is commercially available as an active dry yeast and is used worldwide to improve fermentation processes and wine quality. Some argue that the widespread use of commercial yeasts in recent years has led to the creation of wines that lack character, i.e., sensory characteristics of typical regions of a country. *S. cerevisiae* could act as a biocontrol agent that produces lethal toxins during the winemaking process. The killer strains are able to inhibit the growth of spoilage microorganisms that affect the quality of the wine. The addition of sulfur dioxide as an antioxidant and antimicrobial additive is a common practice in winemaking. This particular strain can convert sulfuric anhydride into sulfur ions, which is undesirable.

5.1.2.25 *Saccharomyces ludwigii*

Saccharomycodes ludwigii is a yeast species known to be a contaminant in alcohol and fruit juice production. It is highly resistant to typical environmental stressors such as high temperature, high sugar concentration, and high sulfur dioxide concentration. It is often referred to as "the winemaker's nightmare" as it contaminates products by overpowering desirable yeast species. *Saccharomycodes ludwigii* commonly contaminates bottled wines and fruit juices. This particular strain creates a significant amount of ethyl acetate.

5.1.2.26 *Schizosaccharomyces pombe*

The yeast species *Schizosaccharomyces pombe* was traditionally considered spoilage yeast. Over the past decade, its use has increased owing to the unique ability of malic acid deacidification to mitigate the strong acidity in wines from Northern Europe. This process converts malic acid into ethanol and CO_2, without generating lactic acid as lactic bacteria do. In recent years, *Schizosaccharomyces pombe* has been increasingly utilized to address challenges in the modern winemaking industry, including the enhancement of food quality and safety. Some novel applications, include the release of high levels of autolytic polysaccharides, reduction of gluconic acid, urease activity which prevents the formation of ethyl carbamate (a toxic compound), increased production of pyruvic acid linked to color enhancement, and the removal of lactic bacteria substrates to prevent the formation of biogenic amines (toxic compounds like histamine) [9]. *Schizosaccharomyces pombe* is rarely present in must and wine, except in cases of must that have undergone heating due to its resistance to it. It also has a high resistance to sulfuric anhydride.

5.1.2.27 *Saccharomyces chevalieri*

Saccharomyces chevalieri is the dominant species of yeast found in coconut palm wine fermentations. It is similar to *Saccharomyces cerevisiae* but lacks the ability to ferment maltose and has been advanced as an intermediate to the development of *S. cerevisiae*.

5.1.3 Growth Cycle of Yeast

In order to be able to analyze the development cycle of microorganisms in general and of yeasts during winemaking in particular, we should know two basic definitions. Growth rate is called the number of divisions in the unit of time. Also proliferation time is called the time necessary for the cell to divide into two cells.

The time for the microorganisms to multiply is proportional to the content of the environment in factors that positively affect the growth of the yeasts and disproportionate to the hindering factors of the environment in which the microorganisms are found.

But let's look in detail at the growth cycle of yeasts in grape must. The yeast life cycle has been categorized by Monod in 1942 and Senez in 1968 into the following 6 stages. Their graphical representation is presented in Fig. 5.1.

5.1.4 Yeast Growth Factors

In addition to yeast nutrition during winemaking, which will be discussed at length in the next chapter. There are some physical, chemical, and biological factors that under conditions positively or negatively affect the growth of yeast cells and increase their fermentation efficiency. There are also a number of other physical, chemical, and biological factors that only negatively affect the growth and effectiveness of yeasts. In order to carry out a smooth alcoholic fermentation, we are obliged with analyzes to be made on the grape skins, to have a complete picture of these physical, chemical, and biological factors and to intervene or not to create the best conditions required to do the job successfully us. These factors are as follows:

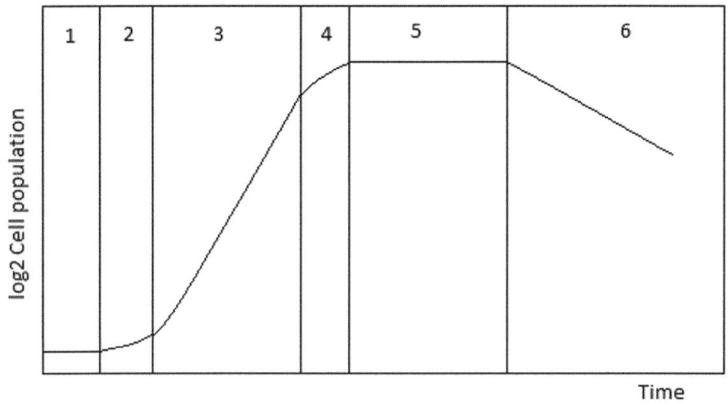

Fig. 5.1 The development phases of yeasts according to Senez, 1968 and Monod, 1942. (1) Latent phase. In this phase the yeast cells adapt to the new conditions found and their growth rate is zero. (2) Acceleration phase. In this phase the proliferation of cells begins and in the diagram it is depicted with the beginning of the rise of the curve. (3) Logarithmic phase. In this phase we have the maximum growth rate of yeast cells with almost zero mortality. (4) Deceleration phase. The growth rate of the yeasts decreases considerably and the curve begins to level off. (5) Stagnation phase. In this phase the growth rate of the yeasts is zero. The curve remains horizontal and the population remains constant for quite some time. (6) Reduction phase. The yeast population decreases and the cells autolyse. The curve takes a downward course

(a) Oxygen
(b) Temperature
(c) pH
(d) Sugar concentration
(e) Fermentation inhibitors
(f) Type of fermentation (Controlled—spontaneous)
(g) Type of vinification

5.1.4.1 Oxygen

Depending on the amount of oxygen available to them, yeasts develop two different activities, fermentation and respiration. In an anaerobic environment, sugars are metabolized into alcohol, carbon dioxide, and numerous other by-products. During this process, 40 kcal of heat is released. In the other case, i.e., in the case where we have the presence of oxygen, the sugars decompose into carbon dioxide and H_2O and for each of their molecules, heat of 686 kcal is produced, energy necessary for the smooth functioning of the yeasts. However, in order to have alcoholic fermentation that takes place in an anaerobic environment, a significant population of yeasts is required, whose multiplication requires, among other things, the presence of oxygen. From the above we conclude that these two phenomena coexist during alcoholic fermentation. This is evidenced by the conditions that prevail in practice before and during alcoholic fermentation. Simply put, there can be no absolute separation between respiration and fermentation or aerobic and anaerobic environments. These two coexist and what changes is essentially the intensity of these two phenomena as a function of time. That is, when the intensity of respiration is low, then the intensity of fermentation is high and vice versa. It is important for the winemaking process to know the ideal moment when we must enhance the phenomenon of cellular respiration to help the development of the yeasts and in this way ensure a smooth alcoholic fermentation. Experiments that have been done have concluded that aeration of the grape must on the second day of vinification gives the highest number of cells. This is due to the fact that on the second day of fermentation the yeasts have adapted to their new environment and the acceleration phase begins. It is worth noting that yeasts do their best work when the redox potential is around 400 mV. This value coincides with the start of alcoholic fermentation. After about 3 days the redox potential drops to 100–150 mV. In order to continue and complete the fermentation smoothly, the redox potential must return to the original levels. This is done by enriching the must with oxygen by stirring and recirculating the grape must in the fermentation tank [10].

5.1.4.2 Temperature

Fermentation temperature has a double technological interest during the winemaking process. In the first case, the fermentation temperature affects the result quantitatively, i.e., how many sugars will be fermented and how much alcohol will be produced at the end. In the second case, the temperature of the fermentation affects the result qualitatively, which means what secondary by-products arise during the alcoholic fermentation that will have a positive organoleptic effect on the final product. But let's start things from the beginning. Exile yeasts are psychrophilic and mesophilic organisms. They can grow and function at temperatures below 20 °C as well as at temperatures up to 40–45 °C. This depends on the fermentation strain that will prevail and conduct the alcoholic fermentation. For each strain there is an optimum temperature.

A general rule is that alcoholic fermentation develops faster the higher the temperature, with a maximum, depending on the strain, of 30–39 °C. In the above rule we must include an additional rule, according to which the higher the temperature, the more limited the fermentation, i.e., the less sugars are fermented and, consequently, an increasingly smaller percentage of alcohol is produced. When the vinification is carried out in a controlled manner with a known yeast strain, the optimal fermentation temperature is known and can be adjusted using a cooling system in the fermentation tank. In the cases of fermentations with native yeasts, the optimal temperature is not known so we apply a temperature definitely below 20 °C at which it has been proven that we have positive results.

5.1.4.3 pH

A pH between 4 and 6 is considered ideal for yeast growth. In musts, however, the pH varies in the vast majority from 2.8 to 3.9. Therefore with regard solely to the variable of pH in relation to the rate of growth of yeasts, the highest level of pH in them is desirable. At the same time, from studies that have been done over the years on the maximum pH levels of the musts, we also have the formation of secondary products which is important for the positive organoleptic evaluation of the wines. At this point we must take into account the fact that a higher pH means a more favorable environment for the growth of unwanted microorganisms [10].

The course of alcoholic fermentation is a fact that it is a multifactorial function. It is a fact that decisions must be made in a short period of time which affect the final result of the fermentation. The golden ratio must be found between the factors that influence this function and of course we must take into account the available equipment of the respective winery as well as the condition of the raw material. Regarding the two factors temperature and pH, analyzed so far, it has been studied that a combination of temperature 18–20 °C and pH 3.30–3.45 creates a very favorable environment for the growth, multiplication and good functioning of yeasts.

5.1.4.4 Sugar Concentration

A factor that significantly affects the growth and metabolism of yeasts is the initial concentration of sugars in the must. More specifically extreme values of sugar concentrations both up and down are not conducive to the growth rate as well as the efficiency of the fermentation. Specifically for sugar concentrations greater than 300 g/L they disrupt the metabolism of the yeasts and we have a longer latent and acceleration phase, reduced fermentation speed and a reduction in the amount of sugars that are finally fermented. To avoid unpleasant situations during alcoholic fermentation, it is important to choose a yeast strain suitable for the type of winemaking we aim to carry out. For example, if we intend to create a wine with a high alcohol content, we must choose a yeast strain that has the ability to withstand and work effectively in high concentrations of sugars and alcohol.

5.1.4.5 Fermentation Inhibitors

Inhibitors of the growth rate of yeasts and consequently of alcoholic fermentation include various factors that are either exogenous or endogenous to the must. The first case refers to factors whose presence in the must is the result of human intervention and the second case refers to factors that either result from the biochemical action of the microorganisms found in the must or are natural components of the grape and the must.

5.1.4.5.1 Exogenous Fermentation Inhibitors

Tannins

Tannins may have a dual origin in must. More specifically, apart from their natural origin which is a natural component of the grape, tannins may be added as an oenological additive to avoid oxidations. Tannins inhibit the growth of yeasts and this leads to a chain reduction of their final effectiveness. The strain Saccharomyces cerevisiae, which is the most common strain in the wine industry, can grow even in an environment containing 3% tannins.

Ethanol

The addition of ethanol as an oenological practice is very common either to increase the final alcoholic strength in low-grade wines or to produce sweet wines. The addition of ethanol inhibits the growth rate of the yeasts. This is important to know in order to know at which stage of the vinification the addition should be made. The production of Mistela (Vin de liquer) is also based on this inhibitory property of alcohol.

Sulfuric Dioxide

The addition of sulfur dioxide is one of the most important inhibitors of exogenous origin. Its inhibitory action is due to the binding of acetaldehyde. The addition of sulfur dioxide does not have an immediate effect and even after its addition some glycolysis takes place at very slow rates. To make sure that we have achieved a correct intervention, the addition of sulfuric anhydride must be accompanied by sterile filtering to ensure the removal of yeasts from our wine.

Natamycin

Natamycin is a natural antimicrobial peptide produced by strains of *Streptomyces natalensis*. It serves as an effective antifungal preservative in a variety of food products, including yogurt, sausages, juices, and wines. Moreover, it is utilized as a bio preservative and is acknowledged as a safe ingredient for diverse food applications [11].

Sorbic Acid

Sorbic acid is generally used to inhibit *Saccharomyces*, which it is quite good at doing. A problem arises when lactic acid bacteria, particularly Oenococcus, esterify sorbic acid to sorbyl alcohol. The sorbic alcohol in turn rearranges to 2-ethoxyhexa-3,5-diene, which emits a strong unpleasant geranium-like odor with an olfactory threshold of about 100 ng/L. The active dose of sorbic acid in wine is 100–200 mg/L and in must is 500–2000 mg/L.

Pesticides

Residues of pesticides in grapes may carry over to the fermentation system during wine production, potentially disrupting the normal growth of Saccharomyces cerevisiae, which could impact the safety and quality of the wine. The precise nature of the interaction between pesticides and *Saccharomyces cerevisiae* remains largely unclear [12].

5.1.4.5.2 Endogenous Fermentation Inhibitors

Gray Mold

The presence of *Botrytis cinerea* can influence the fermentation process. The mold can alter the sugar composition of the must, additional contain compounds that are toxic which can affect the activity of *S. cerevisiae*.

Tannins

As above-mentioned tannins may have a dual origin in must. In this case we are referring to the tannins that have a natural origin from the grape. Especially in red vinifications where we have a long time of coexistence of the stems with the must, their tannin content is increased compared to red and white vinifications. And in this case tannins inhibit the growth of yeasts and this leads to a chain reduction of their final effectiveness.

Ethanol

Ethanol is the main product produced during alcoholic fermentation. A large amount of ethanol results in the cessation of alcoholic fermentation since the yeasts can only work up to a certain percentage of alcohol. This percentage depends on the strain of yeast that has prevailed. Ethanol causes the dissolution of yeast lipids and thus alters their enzyme systems. In simple terms one could say that yeasts produce a product which is lethal to them.

Higher Alcohol

Higher alcohols are toxic for yeast. Specifically, the more carbon atoms are contained in their molecule, the more the functioning of the yeasts is hindered.

Carbon Dioxide

Carbon dioxide as a product produced during alcoholic fermentation also has a small inhibitory effect.

Acetic Acid

Acetic acid, in turn, has some inhibitory effect on the functioning of yeasts. Its presence alone is not a hindering factor, but concentrations greater than 0.3 g/L may create complications in the smooth conduct of alcoholic fermentation.

Killer Effect

"Killer effect" of *Saccharomyces cerevisiae* is its capacity to generate and release toxic proteins, termed killer toxins. These toxins are fatal to other vulnerable yeast cells, usually of identical or closely related species. Killer toxins generally operate by forming pores in the cell membranes of the target cells, causing their demise. The yeast cells that secrete these toxins are resistant to them due to particular immunity mechanisms. This killer effect has practical uses, including the suppression of unwanted yeast strains in the wine production process [13].

5.1.4.6 Type of Fermentation (Controlled: Spontaneous)

The type of alcoholic fermentation we choose to perform plays an important role in the development of the yeasts. There are two options in this case, in the first option we choose the strain we wish to carry out the alcoholic fermentation (Controlled fermentation) in the second case we let nature choose which strain will prevail (Spontaneous fermentation). There is a list of pros and cons for both cases of fermentation. The final choice is a choice made by each winemaker. However, in relation to the efficiency of the fermentation, the case of controlled fermentation is significantly superior since firstly we know in advance the strain that will prevail and secondly we inoculate the gape must with a large number of yeasts and thus ensure that they have developed such a population of yeasts capable of completing the alcoholic fermentation.

5.1.4.7 Type of Vinification

The type of vinification itself has an important role in the growth rate of the yeasts. This is because different types of vinification lead to a different composition of the must and in turn the growth rate of the yeasts is affected as a different composition of the must means that one of the above endogenous or exogenous factors changes. For example, let's say we have a red variety of grape and we will carry out spontaneous fermentation. Half of the quantity is intended for the creation of a red wine and the rest for the creation of a red wine. It goes without saying that the composition of the yeasts that will arise in both cases from the same raw material will create two completely different environments to which the same yeast strain will try to adapt. So we conclude that the type of vinification we choose has an important role in the growth rate of the yeasts.

5.2 Fermentation Activators (Growth Factors)

This category of additives includes chemical compounds that are necessary even in trace amounts for their operation and development. These are two categories of chemical compounds, steroids and vitamins that belong to the B group.

5.2.1 Vitamins

Vitamins are components of enzymes and are essential compounds for yeast growth. In general, yeasts have the ability to synthesize the necessary ingredients for their function. However, these essential components are not always produced in sufficient quantities and yeasts struggle to function effectively. This in some cases results in the fermentations not taking place completely or the fermentation time taking too long. Based on the above and after laboratory measurements, it is deemed necessary to add vitamins that will favor the smooth conduct of the alcoholic fermentation. Each yeast strain has different vitamin needs, both qualitatively and quantitatively. In general, however, the vitamins that are directly related to the good functioning of yeasts are:

5.2.1.1 Thiamine (Vitamin B1)

Thiamine, known as vitamin B1, is an essential nutrient in the fermentation process of wine must, crucial for yeast growth and metabolism. It serves as a cofactor for various enzymes in metabolic pathways, including glycolysis and the tricarboxylic acid cycle, essential for converting sugars into alcohol. Sufficient thiamine levels prevent stuck fermentations, where yeast activity halts prematurely, ensuring a thorough and efficient fermentation process. Thiamine also affects the production of flavor compounds and aids in breaking down amino acids that yield specific aromatic compounds, enhancing the wine's sensory characteristics. Additionally, thiamine offers stress protection to yeast, functioning as an antioxidant to protect yeast cells from oxidative damage during fermentation. The addition of thiamine is done so that we have a reduction in ketone derivatives which in turn bind the sulfite anhydride. This means that we can use smaller amounts of sulfite anhydride after adding thiamine (Technological interest regarding the total content of sulfite anhydride related to the permissible limits allowed by the legislation). It also means that since a smaller amount of sulfite anhydride is bound we have more free sulfite which is essentially the active form of sulfite anhydride. It is important to know that in order to achieve the result we expect, the addition of thiamine to the grape must, must be done immediately after the intense start of the alcoholic fermentation.

5.2.1.2 Pantothenic Acid (Vitamin B5)

Pantothenic acid (vitamin B5) is a B vitamin and an essential nutrient. Pantothenic acid serves as a precursor to coenzyme A (CoA), crucial for multiple metabolic processes in yeast, such as fatty acid synthesis and oxidation. These functions are vital for preserving the integrity of the yeast cell membrane and ensuring overall cellular health [14]. Sufficient pantothenic acid levels can improve fermentation efficiency. It aids the yeast in metabolizing sugars and producing ethanol, ensuring a smooth and thorough fermentation process. Pantothenic acid also affects the generation of volatile sulfur compounds during fermentation, which can alter the wine's aroma and taste. Maintaining appropriate pantothenic acid levels helps regulate these compounds' production, minimizing the chance of undesirable flavors. Thus, pantothenic acid is vital for yeast health and effective fermentation, significantly enhancing the wine's overall quality and contributing to the desired flavor profile and aroma, making it an essential nutrient in the art of winemaking.

5.2.1.3 Nicotinic Acid (Vitamin B3)

Nicotinic acid, also known as vitamin B3, is crucial in the wine must fermentation process. During alcoholic fermentation, especially with *Saccharomyces cerevisiae* yeast, nicotinic acid is vital for maintaining cellular redox balance. This balance is essential for effective fermentation and the creation of desired metabolites. Nicotinic acid serves as a precursor to NAD^+ (nicotinamide adenine dinucleotide), an essential redox cofactor. NAD^+ plays a role in many metabolic processes, including fermentation. Sufficient nicotinic acid levels are vital for the proper regeneration of NAD^+, which is key to maintaining redox balance.

The presence of nicotinic acid significantly affects the rate and thoroughness of fermentation. A deficiency in nicotinic acid may result in slower fermentation and diminished biomass yield, potentially hindering the fermentation process. Nicotinic acid concentrations can alter the variety and quantity of metabolites generated during fermentation, influencing the taste, scent, and overall quality of the resulting wine. Grasping the intricacies of nicotinic acid absorption and its effect on yeast metabolism is crucial for vintners aiming to refine fermentation conditions and enhance the quality of their wine [15].

5.2.1.4 Pyridoxine (Vitamin B6)

Pyridoxine, also known as vitamin B6, plays an important role in the fermentation process of wine must. Specifically supporting yeast health and enhancing fermentation efficiency, which in turn contributes to the wine's overall quality. Effective nutrient management, particularly maintaining sufficient levels of pyridoxine, is key to producing wine that possesses the desired flavors and aromas. The presence of pyridoxine can affect both the rate and the completeness of fermentation. A lack of pyridoxine may result in slower fermentation rates and could lead to stuck fermentations, where the yeast is unable to completely convert sugars into alcohol. Furthermore, Pyridoxine is an essential nutrient for yeast metabolism, serving as a coenzyme in a range of enzymatic reactions. These include the metabolism of amino acids and the synthesis of neurotransmitters. Ensuring sufficient levels of pyridoxine is crucial for the efficient functioning of these metabolic processes in yeast cells. Finally Pyridoxine levels can also impact the production of hydrogen sulfide (H_2S) during fermentation. Low levels of pyridoxine, along with other B vitamins like thiamine, have been associated with increased H_2S production, which can lead to off-flavors in the wine [16].

5.2.1.5 Mesoinositol

Mesoinositol, or myo-inositol, is involved in the fermentation process of wine must. It can affect yeast metabolism and the overall fermentation. Mesoinositol acts as a nutrient for yeast, aiding their growth and activity during fermentation [17]. It can also enhance the efficiency of fermentation by improving yeast performance, leading to a more thorough conversion of sugars into alcohol. Additionally, mesoinositol can influence the wine's flavor profile, aiding in the creation of specific aromatic compounds.

5.2.1.6 Biotin (Vitamin B7)

Biotin, also referred to as vitamin B7, is vital in the fermentation process of wine must. Primarily, biotin is indispensable for yeast metabolism, serving as a coenzyme in several metabolic reactions, aiding yeast cells in the efficient conversion of sugars into alcohol and carbon dioxide. Sufficient biotin levels can prevent stuck or sluggish fermentations, characterized by a decrease or halt in yeast activity before sugar conversion is complete. Additionally, biotin enhances the overall vitality and function of yeast, which can affect the evolution of flavors and aromas in wine.

5.2.1.7 Riboflavin (Vitamin B2)

Riboflavin, also known as vitamin B2, is crucial in the fermentation of wine must, mainly by aiding yeast metabolism. It is vital for yeast growth, serving as a precursor to flavin coenzymes such as flavin mononucleotide (FMN) and flavin adenine dinucleotide (FAD), which participate in various redox reactions essential for energy production and metabolic activities in yeast. Additionally, riboflavin's photosensitivity can lead to the development of off-flavors in wine when exposed to light, in a phenomenon referred to as light-struck taste. This happens when riboflavin interacts with methionine, an amino acid, causing unwanted alterations in the wine's flavor profile [16].

5.2.1.8 Folic Acid (Vitamin B9)

Folic acid, also known as vitamin B9, is a vital nutrient in the fermentation of wine must. It is essential for the synthesis of nucleotides and amino acids, which are necessary for the division and growth of yeast cells. Yeast cells consume folic acid to aid their swift multiplication and metabolic functions during fermentation. Sufficient folic acid levels in the must contribute to a robust and effective fermentation process, enhancing the quality of the wine. Conversely, inadequate folic acid can result in slow or halted fermentations, with yeast unable to efficiently transform sugars into alcohol.

5.2.2 Steroids

A steroid is an organic molecule characterized by four interconnected rings, labeled A, B, C, and D, in a distinct molecular configuration. Steroids serve two primary biological roles: they are vital elements of cell membranes that modify membrane fluidity and function as signaling molecules [18].

5.2.2.1 Sterols

Sterols are vital in the alcoholic fermentation process of wine, playing a key role in preserving yeast cell membrane structure and function. Yeast cells, like Saccharomyces cerevisiae, depend on ergosterol, synthesized under aerobic conditions, and phytosterols, derived from grape musts when oxygen is absent. These sterols enable yeast cells to endure challenging conditions such as ethanol stress and sterol scarcity, preventing slow or stalled fermentations. Ensuring the integrity of cell membranes, sterols help maintain yeast viability and activity during fermentation, which is crucial for the wine's quality and stability [19].

5.2.2.2 Oleic Acid

Oleic acid, an unsaturated fatty acid, is significant in the alcoholic fermentation of wine. It preserves the integrity and fluidity of yeast cell membranes, essential for the survival and function of yeast during fermentation. This becomes particularly critical under stress, such as high ethanol levels. Adding oleic acid to the fermentation process can enhance the yeast's resistance to ethanol's toxicity, thus boosting fermentation efficiency and stability, leading to more reliable and superior wine production [20].

5.3 Fermentation Nutrients

As living organisms, yeasts need food to function effectively and grow satisfactorily. This means that there must be enough and suitable nutrients in the substrate they work on. Grape must contains the necessary nutrients for their nutrition. However, the mass production of wine and modern oenological practices require the fortification of the must in nutrients since it has been observed that toward the end of the alcoholic fermentation these are exhausted. The nutrients necessary for the smooth conduct of alcoholic fermentation could be categorized into the following general groups: Carbon sources, nitrogen sources, and inorganic components.

5.3.1 Carbon Sources

The carbon requirements of yeasts depend on the stage of fermentation as they are directly related to the presence or absence of oxygen.

Specifically, in an environment in the presence of oxygen, the yeast can feed on a large number of compounds such as sugars, amino acids, glycerol, alcohol, etc. The molecules of these carbon compounds as products of yeast nutrition finally give CO_2 and H_2O.

In an anaerobic environment, yeasts use only specific sugars for nutrition and mainly hexoses (depends on the yeast strain), D Glucose, D Fructose, D mannose, and D galactose. They can also use maltose, which with some biochemical functions is broken down into two glucose molecules, and sucrose, which is broken down into glucose and fructose. In this case of the anaerobic environment, this feeding of the yeasts results in the production of alcohol.

5.3.2 Nitrogen Sources

In order for the yeasts to succeed in forming their amino acids as well as multiplying, they have a great need for ammonium nitrogen. Ammonium nitrogen is the most digestible form of nitrogenous ingredients and this results in it being depleted very quickly. When the ammonium nitrogen is depleted, the yeasts draw the necessary source of nitrogen from the amino acids, polypeptides and certain forms of proteins present in the must. Apart from their multiplication, their fermentation capacity also depends on the available nitrogen. From the above we understand how important our intervention in ammonium nitrogen is to achieve effective fermentations. Its addition must always be made before the start of alcoholic fermentation [21].

5.3.3 Inorganic Compounds

Research in recent years has re-examined the mineral requirements of the substrate for the smooth conduct of alcoholic fermentations.

Magnesium is crucial for numerous enzyme functions, including those involved in phosphate transport. Without magnesium, ATP synthesis is impossible, making it vital for glycolytic metabolism. Fermentation media must include bioavailable magnesium. Consequently, adding extra magnesium sources, such as magnesium sulfate, to industrial fermentation media could positively impact alcohol production yields.

It is crucial to recognize that excessive calcium in fermentation media may inhibit vital magnesium-dependent enzyme activities, adversely affecting glycolytic metabolism and impeding fermentation. Calcium competes biochemically with magnesium, which is why a high Mg: Ca concentration ratio is advised for industrial media used in alcoholic fermentations. Magnesium has also been shown to improve yeast stress tolerance by exerting a protective effect on the functional integrity of the cell membrane.

Regarding zinc, deficiency of this essential trace element, at levels <0.1 ppm, can lead to stuck fermentations. This issue has occasionally been observed in fermentation processes, where the final enzyme, alcohol dehydrogenase, is a zinc-dependent metalloenzyme.

The presence of toxic metals should be avoided because even trace amounts of metals such as silver, arsenic, barium, cesium, cadmium, mercury, lithium, and lead can be lethal to yeast cells.

It is also important to consider other mineral nutrients. *S. cerevisiae* requires sulfur to synthesize sulfur-containing amino acids (such as methionine, cysteine, and cystine) from sulfates. Phosphorus is needed for the synthesis of nucleic acids, phospholipids, and ATP, and can be supplied in the form of phosphates to the media. For instance, diammonium phosphate not only provides phosphorus but also additional assimilable nitrogen.

5.4 Malolactic Fermentation Additives

Malolactic fermentation is a secondary fermentation process in winemaking where lactic acid bacteria convert the unstable malic acid into more stable lactic acid, enhancing the wine's flavor and texture. Malolactic fermentation takes place mainly after the end of alcoholic fermentation.

If after laboratory tests we find that it does not start spontaneously, we can intervene in various ways. The simplest way is to mix wine that is in the process of malolactic fermentation with the wine we want to make at a rate of about 30–40%. In any case, we do not sulfurize our wine because this prevents malolactic fermentation from taking place. The temperature should not exceed 20–22 °C and the preservation of the wine that we wish to undergo malolactic fermentation is preferable to be in a large tank rather than divided into smaller ones.

The most effective way is to inoculate the wine with cultures of malolactic bacteria. The most common strains for carrying out malolactic fermentation are *Oenococcus oeni*, *Lactobacillus*, and *Pediococcus*. There are also special organic malolactic nutrients, which improve the growth conditions for bacteria. Also adding enzymes such as lysozyme can be used to inhibit unwanted bacterial strains and control malolactic fermentation [22].

5.5 Clarification and Extraction Enzymes

Enzymes used for clarity of must extraction are pivotal in winemaking, as they enhance the retrieval of essential compounds from grape skins, including aroma precursors, color, and phenolic compounds. These enzymes aid in yielding more juice from grapes, thus streamlining the process. Additionally, by breaking down the cell walls, they release more color and aroma compounds, improving the wine's sensory attributes. Extraction enzymes also shorten the duration required for maceration, settling, and filtration, thereby accelerating the winemaking process. Typically, these enzymes are introduced during the pre-fermentation phase to optimize their impact. The specific dosage and enzyme variant depend on the wine type (red or white) and the sought-after qualities.

5.5.1 Pectic Enzymes

These enzymes, like pectinase, break down pectin—a polysaccharide in fruit cell walls that can make wine cloudy. They are essential for clarifying wine and enhancing juice extraction from fruits. Pectic enzymes are particularly important in winemaking with high-pectin fruits. The optimal time to add coagulant enzymes is right after crushing the fruit and before pressing, which helps release more juice from the fruit fibers. If this step is missed, it's still possible to add them at the start of fermentation. Besides clarifying wine, lactic enzymes can also improve color and juice extraction, leading to enhanced performance and potentially more vibrant wines. For best results, introduce coagulation enzymes at temperatures around 27 °C (80 °F), as lower temperatures may decelerate the reaction.

5.5.2 Proteolytic Enzymes

Proteolytic enzymes, also known as proteases, are essential in winemaking for breaking down proteins that may cause cloudiness and instability. Commonly used to enhance the final product's clarity and stability, these enzymes hydrolyze proteins into smaller peptides and amino acids, aiding in the reduction of cloudiness in wine. This process is especially vital for white wines, where clear protein stability is key. Typically, proteolytic enzymes are introduced during the clarification stage, contributing to the wine's overall quality by improving clarity and stability, leading to a smoother mouthfeel and a more refined flavor profile. In commercial winemaking, proteolytic enzymes are frequently sourced from fungi like Aspergillus niger, cultivated in controlled environments to yield these enzymes [23].

5.5.3 Cellulase and Hemicellulase

Cellulase and hemicellulase are enzymes that assist in breaking down cell walls into simpler sugars, which can enhance the extraction of color and tannins from grape skins, thereby improving the wine's body and flavor. These enzymes are crucial in winemaking, especially for boosting extraction and clarification.

Hemicellulase targets hemicellulose, a component of the cell wall, aiding in the conversion of complex carbohydrates into fermentable sugars. These enzymes are commonly added during the maceration process to improve the extraction of compounds from grape skins, which is especially beneficial in red winemaking for enhancing color and tannin extraction. The breakdown of cell walls by these enzymes releases more juice from the grapes, thus increasing the yield. They also assist in the clarification process by decomposing polysaccharides that cause cloudiness, yielding a clearer wine. Moreover, the cell wall degradation facilitates better extraction of phenolic compounds, enhancing the wine's flavor and color profile. Typically, these enzymes are sourced from microbes like fungi, which are grown in controlled environments to produce the enzymes in bulk [24].

5.5.4 Beta-Glucanase

Beta-glucanase is an enzyme that breaks down beta-glucans, polysaccharides present in yeast cell walls and certain fungi like Botrytis cinerea. These substances can cause issues with wine filtration and clarity. In winemaking, beta-glucanase plays a crucial role in enhancing wine clarity and filterability. It is utilized during the clarification and filtration processes. Additionally, it aids in releasing mannoproteins and oligosaccharides from yeast cell walls, contributing to the protein stability of the wine. Beta-glucanase for commercial use is typically obtained from fungi, such as Trichoderma species, grown in controlled environments to produce the enzyme [25].

5.6 Maturations Enzymes

Enzymes that mature wine are vital for improving the aging process and the final product's quality.

5.6.1 Glycosidases

These enzymes are key in winemaking, especially for enhancing the wine's aroma and flavor profile. They liberate volatile aroma compounds like terpenes, norisoprenoids, and benzene derivatives, enriching the wine's bouquet. Glycosidases aid in developing intricate flavors by breaking down glycosides, thus elevating the wine's sensory appeal. The release of these aromatic compounds adds depth and complexity, making the wine more attractive to consumers [23]. Glycosidases are generally active during fermentation and aging. To maximize the release of aromatic compounds, winemakers might choose specific strains of yeast or bacterial cultures. Many wine yeasts, especially those from non-Saccharomyces species, secrete active glycosidases during fermentation. Oenococcus oeni, a bacterium involved in malolactic fermentation, also produces glycosidases that enhance the wine's aromatic profile. β-Glucosidases are the most extensively researched glycosidases in winemaking. They cleave glycosidic bonds to free aromatic compounds from their glycosidic forerunners. Furthermore α-Rhamnosidases, α-Arabinosidases, and β-Apiosidases also play a role in releasing aroma compounds [26].

5.6.2 Glucanases

Glucanases are enzymes that break down glucans, improving the filtration and stability of wine. They enhance aroma and flavor by decomposing complex molecules, while maturation enzymes release additional aroma and flavor compounds, enriching the wine's sensory qualities. Glucanases and proteases also reduce haze, enhancing the wine's clarity and stability [23]. Furthermore, these enzymes can expedite certain aging processes, allowing the wine to be ready for consumption sooner without compromising its quality. Maturation enzymes are usually added during the aging process, and the specific type and amount depend on the desired wine characteristics and the winemaker's objectives. Glucanases aid in wine stability by preventing the formation of unwanted polysaccharide complexes [25]. Some wine yeasts naturally produce glucanases during fermentation. These enzymes, typically derived from fungi *like Aspergillus niger* and *Trichoderma* species, known for robust enzyme production, are added during fermentation or aging to ensure effective breakdown of glucans, thus facilitating better processing and quality of the final product.

β-1,3-Glucanases are enzymes that break down β-1,3-glucans, polysaccharides present in the cell walls of yeast and some fungi.

β-1,6-Glucanases target β-1,6-glucans, another polysaccharide type that can cause filtration issues in wine.

5.7 Tannins

Tannins, naturally occurring compounds in plants, are found in grape skins, seeds, and stems. They are essential in winemaking, serving several key functions:

Taste and Texture: Tannins contribute to the astringency and bitterness of wine, balancing its sweetness and fruitiness. This astringency also enhances the wine's body and mouthfeel.

Aging Potential: Tannins serve as natural antioxidants, preserving the wine and enabling it to age well. They support the polymerization process, where tannin molecules bond, softening the wine and increasing its complexity.

Color Stabilization: Tannins stabilize the color of red wines by binding with anthocyanins, the pigments responsible for the wine's hue. This binding prevents the pigments from precipitating out of the wine.

Oxidation Prevention: Tannins incorporate oxygen, preventing more damaging oxidative reactions and thus safeguarding the wine from premature oxidation.

Tannins are more pronounced in red wines due to prolonged contact with grape skins and seeds during fermentation. White wines generally have lower levels of tannins as they are typically fermented without skins and seeds.

Tannins during winemaking either come naturally from the grape and are extracted from grape skins, seeds, and stems during fermentation. Either from oak-derived tannins introduced through aging in oak barrels or using oak shavings, coils or dust. Whether added by us as commercial tannin products or special tannin powders or liquids, they can be added at various stages of winemaking to achieve the desired results.

In the first case, it is clear that the content of tannins in our wines depends mainly on the type of vinification we have chosen.

In the second case, it is a bit more complicated as the type of barrel and the contact time of our wine with it has a direct relationship with the future tannicity and organoleptic character of our wine. Also in the same category have been created aging chips which give the wine the characteristics of a wine that has gone through a barrel without actually going through a barrel. And in this case of aging chips, the type of chips and the extraction time are very important and we must consult the manufacturers of the chips to choose the right ones for the desired type of wine we plan to make.

In the last case where we choose to use additional tannin powders or liquids, our purpose is either to protect the wine from any oxidation (mainly in white and rosé wines) or to improve the organoleptic quality of our final wine. Choosing the right tannins is a process that requires a lot of testing and a lot of expertise in order to choose the right tannins that will organoleptically improve our final wine.

In the last case of adding tannins as an oenological additive in powder or liquid form there are many options. The most common ones are detailed below:

1. Powdered tannins derived from oak heart tannin that improve organoleptic characteristics and impart vanilla notes.
2. Oak gallotannin in fermentation powder with an antioxidant effect that prevents reducing odors and protects in cases where we have grapes affected by Botrytis.
3. Mixture of gallotannins, ellagic tannins, and procyanidic tannins in powder for vinification of red wines and provide antioxidant protection and color stabilization.
4. Powdered oak gallotannin for white wines that provides protection against oxidation during alcoholic fermentation.
5. Powdered oak tannin with pyrocatechin for red wines which provides color stabilization.
6. Simple oenological tannin powder for antioxidant protection of wines during fermentation and maturation.
7. Powdered grape tannin to improve the structure of white wines added before bottling.
8. Powdered grape tannin to stabilize color and improve structure added during fermentation.
9. Powdered grape tannin for white and red wines added before bottling.
10. Powdered grape tannin to stabilize color and improve structure.
11. Powdered oak tannin for added complexity with hints of caramel, vanilla and mocha.
12. Oak tannin powder to improve structure with notes of coconut and vanilla.
13. Powder tannin added for revitalization and fizzing before bottling and imparts aromatic clarity.
14. Liquid tannin for color stabilization of red wines that provides antioxidant protection for white and red must.
15. Liquid oak tannin for structure, complexity and finesse in white, rosé and red wines.
16. Liquid tannin that imparts citrus aromas to white and rosé wines.
17. Liquid or powdered tannin to protect white and rosé wines from oxidation.

5.8 Bentonites

Bentonite is vital to winemaking, serving two main purposes: clarifying and stabilizing proteins.

To clarify, bentonite helps remove particles that contribute to wine cloudiness. It clings to these particles, facilitating their descent to the bottom, which clears the wine.

In terms of protein stabilization, bentonite proves effective in extracting proteins that can lead to cloudiness in wine, especially white varieties. It attaches to these proteins, forming aggregates that are easily eliminated.

The addition of bentonite can be done either in must during alcoholic fermentation or in wine. When we add bentonite during alcoholic fermentation, we have a double benefit since its addition helps the desalination process and additionally removes the proteins and a part of the oxidizing enzymes that are of a protein nature. When bentonite is added to pure wine, it is done for reasons of protein stabilization in which we avoid any turbidity due to the proteins.

As a material, bentonite, is an impure clay resulting from the weathering of volcanic ash, is an absorbent material that binds suspended particles, reducing wine cloudiness.

The main types are sodium and calcium bentonite. Each contains trace elements, but they are characterized by the dominant mineral. Both can clarify wine. Their difference has to do with the residuality of the metals after their removal from the wine.

5.8.1 Sodium Bentonite

Natural sodium bentonite, with balanced sodium and calcium ions, expands well during rehydration, offering a large surface area for ion exchange. This feature allows it to bind heat-labile proteins in juice and wine, creating a neutral floc that settles over time.

5.8.2 Calcium Bentonite

Natural calcium bentonite, rich in calcium ions and low in sodium ions, has a mesh structure that flocculates effectively, ideal for wine lees clarification and compression. Its less open structure means a reduced ability to bind proteins.

5.8.3 Bentonite in Mixtures

Bentonite is often combined with other clarifying agents such as PVPP and casein, which are light and settle slowly. Bentonite accelerates their sedimentation, enhancing clarification. Betonite is a fairly dense material and if not properly prepared, it will collect at the bottom of the tank and have very little action. For our intervention to have a satisfactory result, we must prepare the bentonite strictly according to the manufacturer's instructions. However, the hydration of bentonite is an important process. All bentonite must be properly inflated to function as intended. Its swelling is affected by two factors related to water quality, namely water hardness and pH:

5.8.3.1 Water Hardness

Hard water is particularly problematic when calcium-based bentonites are used. Water hardness refers to the level of inorganic ions in the water. The more mineral ions dissolved in the water, the harder the water. Calcium-based bentonites will not expand to maximum capacity if swollen in hard water. If calcium-based bentonite is preferred, then soft (deionized) water should be used. Sodium-based bentonites are little affected by hard water minerals.

5.8.3.2 pH of the Water

If bentonite is prepared in an acidic solution or wine, it immediately sets and loses up to 50% of its effectiveness. It is worth mentioning that now the companies that create oenological additives have created various types of bentonite that can be found either in the classic powder form, or in granular form or even in the form of micro-ionized bentonite for tangential filtration filters. It is also important that the hydration time varies between 30 min and 24 h. Finally, bentonite has been released which can be directly dissolved in wine without the hydration stage.

5.9 Clarification Additives

Vinification, additional clarifications are necessary to eliminate particulate matter and sediment, leading to a cleaner and more stable wine. The document is some common plugins published in this process:

Bentonite: Discussed in the previous section in detail.

Gelatin: Gelatin is widely used in winemaking as a fining agent, especially to soften red wines by reducing tannins and astringency. This agent is derived from animal collagen, often obtained from boiled skin, tendons, ligaments, and bones. Its mechanism involves binding with tannins and other phenolic compounds, causing precipitation and thus clearing the wine and reducing its astringency. Typically, gelatin is introduced into the wine fermentation to form solid clumps, or "flakes," which are then easily removed. However, using too much gelatin can strip red wines of their color and lead to protein instability. Therefore, it is vital to use gelatin sparingly and usually in combination with other thinning agents such as Kieselsol.

Isinglass: Isinglass is a fining agent obtained from the swim bladders of fish, mainly utilized in winemaking for the clarification and stabilization of wine. It aids in the removal of solids and impurities, yielding a wine that is clearer, brighter, and more stable. Isinglass excels at eliminating polyphenolic compounds, which can soften harshness and astringency. Commonly, it is employed as the last step in the clarification process prior to bottling. Isinglass operates by adhering to particles within the wine, causing them to settle and facilitating the separation of the clear wine. In contrast to other fining agents, isinglass has a negligible effect on the body and astringency of the wine, hence it is often the fining agent of choice for white wines.

Casein: Casein, a protein sourced from milk, serves as a fining agent in the winemaking process. Its primary function is to diminish astringency and mellow the tannin profile in white wines. It effectively removes discoloration due to oxidation and can also neutralize excessive oakiness. While it is commonly introduced in a purified state, skim milk is an alternative. Casein operates by attaching to phenolic substances and other particulates, facilitating their removal from the wine. As a mild fining agent, casein preserves the wine's inherent aromas and flavors, making it an ideal choice for fine white wines where maintaining original qualities is crucial.

Kieselsol (Silicon dioxide) and Chitosan: Kieselsol and Chitosan are commonly paired in winemaking to produce a clear and stable wine.

Kieselsol, a fining agent with a negative charge derived from silicon dioxide, is adept at removing positively charged particles like proteins and compounds that cause haze.

Chitosan, sourced from crustacean shells such as shrimp and crabs, carries a positive charge. It is effective in eliminating negatively charged particles, including yeast and phenolic compounds. Together, these agents are efficient in eradicating particles that cause haze, yielding a crystal-clear wine. They enhance the wine's taste and aroma by removing undesirable compounds and excess tannins, resulting in a smoother and more harmonious wine. The combined use of Kieselsol and Chitosan can greatly improve your wine's quality and visual appeal.

Usage Instructions

- Step 1: Introduce Kieselsol to the wine first, which binds the unwanted particles into larger clusters for easy removal. The standard dosage is approximately 1 mL per liter of wine.

Step 2: After Kieselsol has acted for a few hours or overnight, incorporate Chitosan to further purify the wine by absorbing and precipitating the remaining impurities. The suggested dosage is around 4 mL per liter of wine.

5.9.1 Albumin

Albumin, usually found in egg whites, is used in winemaking as a fining agent. This process helps clarify and stabilize the wine by removing unwanted particles and reducing astringency. Albumin binds to suspended particles such as dead yeast cells, grape fragments and other solids, causing them to coagulate and settle to the bottom of the wine barrel. This makes the wine clearer. It is particularly effective in red wines for removing hard tannins, which can make the wine bitter and astringent. By softening these tannins, albumin helps create a smoother, more balanced wine. Albumin is commercially available in powder form. In older or traditional vinifications, 3– 8 egg whites are used for a 225 L barrel of red wine. The egg whites are mixed with a small amount of water and then added to the wine, where they perform their subtle action.

5.10 Stabilization Additives

In winemaking, stabilizing additives are vital to maintaining clarity, stability, and protecting the wine from unwanted secondary fermentations or sedimentation. The most frequently used stabilizing additives will then be analyzed in the random order.

5.10.1 Potassium Sorbate

Potassium sorbate serves to inhibit the reproduction of yeast, thus stabilizing the wine and preventing spoilage, which is vital for the sweetness of the wine before bottling. It is commonly used in winemaking to prevent re-fermentation after bottling, a critical measure for sweet wines that may otherwise undergo unwanted fermentation due to residual sugars. Typically, it is introduced after primary fermentation when the wine is clear or has been decanted. This step ensures that any residual yeast cannot multiply. Often, potassium sorbate is used alongside sulfites, such as potassium metabisulfite, to enhance its preservative action. Sulfites are added to suppress microbial activity and prevent oxidation, contributing to wine stability. To apply potassium sorbate effectively, it should first be dissolved in a moderate amount of hot water or wine, ensuring that it is evenly distributed throughout the wine.

5.10.2 Sulfur Dioxide

Sulfur dioxide (SO_2) is crucial in winemaking for its preservative and antioxidant properties. It serves as a general antiseptic, targeting wine microorganisms, including yeasts and bacteria such as acetic and lactic. It hinders the growth of harmful bacteria and wild yeast, preventing spoilage and ensuring controlled fermentation by preferred yeast strains. SO_2 shields wine from oxidation, which can cause browning and off-flavors. Beyond its antioxidant role, it also combats oxidases, enzymes present in molasses responsible for various oxidation processes, by destroying them.

Adding sulfur dioxide to must safeguards the color of white wine, preventing it from turning yellow due to oxidation. In red wine production, sulfuring the pulp enhances the red hue. By reacting with oxygen, it maintains the wine's color and freshness. Moreover, SO_2 prolongs wine's shelf life by preventing microbial growth and oxidation during storage and aging. Sulfurizing the must also delays the onset of alcoholic fermentation, aids in must clarification, and stabilizes fermentation by moderating yeast activity, which helps keep the temperature at optimal levels.

Sulfur dioxide is available in different forms, such as potassium metabisulfite or sodium metabisulfite, with usage amounts varying according to the wine's pH. It is sold as a powder to be dissolved in water or as a liquid. Typically, wines with lower pH require less SO_2. In white winemaking, must sulfuration should occur immediately after pressing. For red winemaking, it should be added right after crushing the grapes. If sulfur dioxide is introduced after fermentation begins, it loses its protective effect on the must.

Winemakers closely monitor SO_2 levels to keep them in an effective range without overuse, as too much SO_2 can result in off-flavors and odors.

5.10.3 Ascorbic Acid

Ascorbic acid, commonly known as Vitamin C, is a prized additive in winemaking for its antioxidant capabilities. It serves as an antioxidant by neutralizing free radicals that lead to oxidation, thus preventing undesirable flavors, discoloration, and loss of freshness in the wine. Ascorbic acid's reaction with oxygen preserves the wine's vivid color and prevents browning. It also maintains the intricate balance of flavors, ensuring that the wine's fruity, floral, and spicy notes stay vibrant and reflect their true nature. The optimal moment to add ascorbic acid is immediately before bottling to offer the utmost protection against oxidation at this crucial stage. Ascorbic acid is frequently used in conjunction with sulfites to bolster the wine's stability and extend its shelf life.

5.10.4 Potassium Hydrogen Tartrate

Potassium Hydrogen Tartrate, commonly known as cream of tartar, is crucial in winemaking, especially during the cold stabilization process. It is utilized to inhibit the development of tartrate crystals in the wine once bottled. In cold stabilization, the wine is cooled to nearly freezing points, prompting the tartrate crystals to emerge and settle out. This step ensures that crystals do not form in the bottle later, which consumers might find unattractive. It preserves the wine's esthetic appeal by preventing crystals. Enhanced stability contributes to the wine's chemical consistency, ensuring its taste and quality remain uniform over time. It is typically introduced post-fermentation and pre-bottling during the cold stabilization stage, with the quantity varying based on the wine's characteristics and the targeted stabilization level.

5.10.5 Metatartaric Acid

Metatartaric acid serves as an important additive in winemaking, chiefly utilized to stabilize wine by hindering the crystallization of tartrates. It prevents the formation of potassium bitartrate (KHT) and calcium tartrate crystals, which can mar the appearance of bottled wine. Its effectiveness is notable in wines intended for consumption within a 6-month period post-bottling. Additionally, it functions as an acidulant, aiding in the suppression of spoilage organisms, thus offering a cost-effective option for wines with a brief shelf life. It is generally introduced post-fermentation and pre-bottling, with the quantity adjusted based on the wine's specific composition and stabilization needs. For winemakers aiming to maintain clarity and stability in their wines, particularly those destined for rapid market consumption, metatartaric acid is a strategic selection.

5.10.6 Carboxymethylcellulose (CMC)

Carboxymethylcellulose (CMC) is commonly used in winemaking, mainly to stabilize wine and prevent tartrate crystal formation. It inhibits potassium bitartrate crystallization by blocking nucleation sites, thus preventing crystal growth and precipitation. However, CMC can crosslink with wine proteins, potentially causing haze, so it's essential to confirm protein stability before its addition. CMC preserves the wine's clarity and appearance without altering its pH, acidity, or taste. Available in dry or liquid forms, liquid CMC is preferable for large-scale use. The dosage varies with the wine's makeup, and adherence to manufacturer's guidelines and bench trials is advised for determining the right amount. Wine should not be filtered 24–48 h before and 2–5 days after CMC addition to ensure complete integration. CMC does not stabilize calcium tartrate due to different crystal surface charges.

5.10.7 Dimethyl Dicarbonate (DMDC)

Dimethyl Dicarbonate (DMDC), also known as Velcorin, is a vital additive in winemaking, primarily used for its antimicrobial properties. DMDC effectively controls spoilage microorganisms like yeast, bacteria, and molds by penetrating their cell walls and disrupting metabolic processes, resulting in their inactivation. It is often utilized when other sterilization methods, such as pasteurization or sterile filtration, are not viable. DMDC ensures the wine remains stable and free from unwanted microbial activity. Upon breakdown, DMDC yields methanol and carbon dioxide, both naturally present in wine, thus not altering the wine's flavor or aroma. The recommended dosage of DMDC varies with the wine's pH and microbial load, necessitating trials to ascertain the optimal amount for a particular wine. It is efficacious at low concentrations, generally not surpassing 200 ppm, offering a cost-effective solution for winemakers. Administered just before bottling, it is imperative to handle DMDC with care and adhere to safety guidelines due to its reactive nature. Regulatory bodies, including the FDA, approve DMDC for winemaking, classifying it as a processing aid, not a preservative. Proper handling and application are crucial for ensuring safety and efficacy. Winemakers must receive training in its use and comply with all safety protocols. DMDC serves as a potent tool for winemakers striving to produce stable, high-quality wines without affecting taste or aroma.

5.10.8 Gum Arabic

Gum arabic, or gum acacia, is a valuable additive in winemaking, offering multiple benefits. It prevents tartrate crystal formation and cloudiness in wines. Its molecular structure, featuring hydrophilic and hydrophobic elements, enables interactions with wine components like polyphenols and aromatic compounds. It stabilizes anthocyanins in young red wines, preserving their vibrant color. Additionally, gum arabic softens the astringency of tannins for a smoother, fuller-bodied taste. It enhances the wine's body and volume, contributing to a richer, more balanced sensation. It also helps preserve and stabilize the wine's aroma over time. Serving as a protective colloid, it prevents the aggregation and precipitation of colloidal particles. Typically available as a 20% liquid solution, it is added post-fermentation and pre-bottling. Ensuring protein stability in the wine before adding gum arabic is crucial to prevent haze. The usage amount varies based on the wine's characteristics and the intended outcome, with manufacturer guidelines and bench trials recommended to find the optimal dosage. Gum arabic is a natural, effective means for winemakers to improve their wine's stability, mouthfeel, and overall quality.

5.10.9 Polyvinylpolypyrrolidone (PVPP)

Polyvinylpolypyrrolidone (PVPP) is a synthetic polymer extensively utilized in winemaking due to its fining and stabilizing properties. PVPP effectively reduces phenolic compounds, such as catechins and leucoanthocyanins, which can lead to browning and astringency in wine. It aids in removing unwanted color compounds, especially in white wines, by binding to anthocyanins and other phenolics. PVPP also helps prevent oxidative spoilage by removing these compounds, thereby enhancing the wine's clarity and visual appeal. It contributes to the wine's overall stability, ensuring consistent quality over time. PVPP is inert, leaving no residual taste or odor in the wine.

In application, PVPP is available in powder form and is typically used as a slurry. It can be introduced at various stages, such as before pre-bottling filtration, during or after tartaric stabilization, or following initial clarification. The quantity used varies based on the wine's composition and the desired outcome. Bench trials are recommended to determine the optimal dosage. Post-treatment, PVPP must be removed through filtration, usually with filter media ranging from 3 to 5 microns. PVPP is commonly used alongside other fining agents like bentonite or casein to increase effectiveness. PVPP stands as a versatile and efficient tool for winemakers who strive to produce clear, stable, and superior-quality wines.

5.10.10 Chitosan

Chitosan, a natural polysaccharide obtained from chitin found in crustacean exoskeletons and fungal cell walls, is highly regarded in winemaking for its array of beneficial properties. It combats spoilage microorganisms such as *Brettanomyces*, *Lactobacillus*, *Pediococcus*, and *Acetobacter* by disrupting their cell membranes, leading to their inactivation. In the clarification process, chitosan binds with and precipitates unwanted particles like proteins and polyphenols that can cause wine haze. It also chelates heavy metals such as copper and iron, preventing them from catalyzing oxidation reactions that could spoil the wine. Recognized as safe and approved for organic winemaking, chitosan leaves no residual taste or odor, preserving the wine's sensory qualities. It is applicable to both red and white wines and is effective at various winemaking stages. Available in powder form, chitosan is typically dissolved in water and added to wine post-fermentation or pre-bottling. The dosage varies based on the wine's composition and the targeted issue, with bench trials recommended to determine the optimal amount. Chitosan is a versatile and indispensable tool for winemakers dedicated to crafting clear, stable, and superior wines.

5.11 Polysaccharides

Polysaccharides significantly affect winemaking, affecting both the technological and sensory aspects of wine. They stabilize wine by inhibiting protein cloudiness and tartar crystal formation. In addition, polysaccharides improve the body and texture of the wine, giving it a fuller, more balanced feel. They can also bind to aromatic compounds, helping to preserve and stabilize them. Polysaccharides are naturally released during fermentation, particularly through yeast autolysis. Winemakers may introduce commercial polysaccharide products to increase certain wine characteristics. However, it is important to confirm the protein stability of the wine before adding polysaccharides to avoid cloudiness. The specific amount and type of polysaccharide used may vary depending on the intended effect and composition of the wine, and bench testing is recommended to determine the ideal dosage. Polysaccharides serve as an essential tool for winemakers striving to create superior wines with improved stability, mouthfeel and sensory properties.

5.11.1 Types of Polysaccharides

5.11.1.1 Mannoproteins

Mannoproteins, a type of polysaccharide found in yeast cell walls, especially in *Saccharomyces cerevisiae*, are crucial in winemaking due to their multiple benefits. They stabilize potassium bitartrate in wine, diminish the necessity for cold stabilization, and prevent the formation of crystal nuclei, thus averting tartrate crystal development. Additionally, they enhance protein stability, which helps to avoid haze in wine. In red wines, mannoproteins interact with anthocyanins—the pigments that give red wine its color—to stabilize it. They also improve the mouthfeel of wine, giving it a fuller and smoother sensation. Mannoproteins can lessen the astringency and bitterness of tannins, especially in red wines, and aid in retaining and stabilizing the wine's aroma, thereby improving its sensory profile. Mannoproteins are released naturally during yeast autolysis, particularly when wines are aged on lees. Commercial mannoprotein products are available for addition to wines to attain specific stabilization and sensory outcomes. It is important to ensure the wine is protein-stable before adding mannoproteins to prevent haze formation. The quantity used can vary based on the wine's composition and the desired effect, with bench trials recommended for determining the optimal dosage [27].

5.11.1.2 Arabinogalactans

Arabinogalactans, polysaccharides present in grape cell walls, play a crucial role in winemaking. They aid the clarification process by binding with proteins and other particles for easier removal. These polysaccharides also prevent haze formation, enhancing wine stability and clarity. Arabinogalactans improve the mouthfeel of wine, giving it a smoother and more balanced sensation. They also contribute to the visual appeal and consistent quality of the wine over time by enhancing its body and texture. Released naturally during fermentation and aging on lees, winemakers may add commercial arabinogalactan preparations to improve specific wine properties. It's important to ensure the wine is protein-stable before adding arabinogalactans to prevent haze. The dosage can vary based on the wine's composition and desired effect, with bench trials recommended for determining the optimal amount. Arabinogalactans are an essential tool for winemakers to produce clear, stable, and high-quality wines.

5.11.1.3 Rhamnogalacturonans

Rhamnogalacturonans, complex polysaccharides in grape berry cell walls, play a crucial role in winemaking, especially for stabilization and sensory enhancement. They stabilize wine by interacting with tannins and proteins, reducing reactivity and preventing haze. These polysaccharides also modulate wine aroma by binding volatile compounds, thus enhancing the aromatic profile. They make wine smoother and more palatable by reducing astringency and bitterness through tannin interaction. Rhamnogalacturonans maintain wine clarity and visual appeal by preventing colloidal particle aggregation. They improve the sensory experience by contributing to a fuller, balanced mouthfeel and stabilizing flavor compounds, ensuring consistent taste over time. Released naturally during fermentation and aging on lees, winemakers may also add commercial rhamnogalacturonan preparations to enhance wine properties. It's important to ensure wine is protein-stable before adding rhamnogalacturonans to prevent haze. Dosage varies based on wine composition and desired effects, with bench trials recommended for optimal dosing. Rhamnogalacturonans are invaluable for producing clear, stable, high-quality wines [27].

5.12 Vegetable Proteins

Vegetable proteins are increasingly being used in winemaking as fining agents, offering a plant-based alternative to traditional animal-derived proteins. Vegetable proteins offer an excellent alternative for those allergic to animal proteins such as casein or gelatin. Their use in winemaking promotes sustainable and ethical practices. These proteins effectively bind to and remove undesirable compounds like tannins and phenolics, enhancing the wine's clarity and stability. Typically available in powder form, they are dissolved in water before being added to the wine and can be introduced at different stages, including post-fermentation and pre-bottling. The quantity required varies based on the wine's characteristics and the intended outcome. Conducting bench trials is advisable to ascertain the optimal amount. It's crucial to confirm the wine's protein stability prior to the addition of vegetable proteins to prevent haze development. Winemakers should always verify local regulations concerning the use of particular fining agents. Vegetable proteins are an invaluable asset for winemakers striving to craft clear, stable, and superior wines while committing to ethical and sustainable methods.

5.12.1 Vegetable Proteins Types

Pea Protein: Known for its effectiveness in reducing tannins and phenolic compounds, pea protein is widely utilized for its fining properties.

Potato Protein (Patatin): It is recognized for its capacity to diminish astringency and bitterness, making patatin a favored fining agent.

Soy Protein: Employed for its clarifying properties in wine and the removal of undesirable phenolic compounds.

Wheat Gluten: This plant-based alternative aids in the clarification and stabilization of wine [28].

5.13 Acidity Regulators

The "subtraction" technique for stabilizing tartaric acid in wine reduces the solubility of tartrates, prompting crystallization. This is followed by filtration to remove the crystallized tartrates. This method effectively lowers the concentration of tartrates, securing the wine's stability during storage and distribution. It mainly involves cold stabilization, ion exchange methods, and electrodialysis.

5.13.1 Cold Stabilization

The solubility of potassium hydrogen tartrate (KHT) in wine increases with temperature. As the temperature rises from −4 to 20 °C, KHT's solubility triples. Utilizing this principle, wine can be chilled to 0.5 °C above its freezing point and stirred continuously for one to several weeks, causing tartrate to supersaturate and crystallize. The wine is then filtered at low temperatures to remove KHT crystals. Afterward, the wine's temperature is gradually increased to storage levels to ensure tartaric acid stability by achieving an unsaturated state. This technique involves three cold treatment methods: long-term treatment, contact stability, and continuous stability. The first method avoids adding tartaric crystals, relying solely on prolonged cold treatment for stabilization. In contrast, the other two methods accelerate cold stabilization by adding tartaric crystals during the process.

5.13.2 Ion Exchange Method

Ion exchange treatment involves removing potassium (K^+) and calcium (Ca^{2+}) ions from wine by swapping them with cations in the resin. This method reduces the risk of tartaric acid precipitation by lowering the concentrations of K^+ and Ca^{2+} in the wine. The success of ion exchange in stabilizing tartaric acid depends on the type of resin used and its activation process. Ion exchange resins are divided into anion and cation exchange resins. Research indicates that using cation exchange resins to decrease K^+ and Ca^{2+} levels is more effective for tartaric acid stabilization than using anion exchange resins to reduce HT^- and $T2^-$ levels, and it has less negative impact on the wine.

5.13.3 Electrodialysis

Electrodialysis is a process in which ion exchange membranes selectively allow ions to pass through when subjected to an external direct current. This process causes anions and cations to migrate in opposite directions, traversing the anion and cation exchange membranes, respectively. The goal is to achieve desalination or concentration. During electrodialysis of wine, ions such as HT^-, $T2^-$, K^+, and Ca^{2+} are driven toward the opposite electrodes by the potential difference and exit through the selectively permeable membranes into polar water chambers, leading to the extraction of components responsible for tartrate formation.

5.13.4 Enological Products for Tartaric Stability

The term "tartaric stability in wine" refers to the use of stabilizers to prevent the precipitation of tartrates. Common stabilizers include gum arabic, metatartaric acid, carboxymethyl cellulose, mannoprotein, and potassium polyaspartate.

5.13.4.1 Gum Arabic

Gum arabic, a secretion of Acacia senegal (Linn.) Willd, consists of hydrophobic proteins, macromolecular polysaccharides, and their calcium, magnesium, and potassium salts. It can react with tartrate to form a polymer, creating a hydrophilic layer on the polymer's surface. This enhances the solubility of tartrate in wine, prevents crystal nucleus formation, and preserves the wine's metastable state. Typically used before bottling and filtration, gum arabic is not recommended for wines meant for aging or extended storage, as it prevents the formation of natural sediments, leading to emulsification.

5.13.4.2 Metatartaric Acid

Metatartaric acid, a polyester derived from the esterification of tartaric acid molecules, creates soluble complexes with K^+ and Ca^{2+} ions in wine, thus increasing tartrate solubility. It wraps around small tartrate crystals, preventing their growth and ensuring the wine's stability without precipitation. Typically, metatartaric acid is applied after the final filtration and just before the wine is bottled. It serves as an effective inhibitor of tartrate crystal nucleation, offering a low-cost solution with minimal impact on the wine's characteristics. However, extended storage or exposure to high temperatures can cause metatartaric acid to hydrolyze back into tartaric acid, affecting the wine's tartaric stability. The efficacy of metatartaric acid also varies with the type of wine. For instance, red wine stored at 20 °C for 5 months with 100 mg/L metatartaric acid may become unstable due to hydrolysis, whereas white wine can remain stable for up to 12 months under similar conditions. Therefore, metatartaric acid is more suitable for wines that are in high demand and consumed relatively quickly.

5.13.4.3 Carboxymethyl Cellulose

Carboxymethyl cellulose, a benign polysaccharide that is heat and acid-resistant, possesses a negative charge in wine. It competes with HT^- for K^+ binding, which reduces the concentration of KHT and keeps the tartrate from saturating. Additionally, carboxymethyl cellulose binds to the positively charged surfaces of KHT crystals, reducing the availability of free ions necessary for crystal growth and thus inhibiting the formation of crystals. The additive carboxymethyl cellulose provides a cost-effective and durable solution. White wines treated with 40 mg/L of carboxymethyl cellulose maintain tartaric stability for at least 1 year. Carboxymethyl cellulose is commonly used as a tartaric stabilizer in the production of high-quality and premium wines. However, its addition can alter the color and clarity of red wines. Consequently, the OIV limits the use of carboxymethyl cellulose exclusively to white and sparkling wines. Additionally, carboxymethyl cellulose may interact with wine proteins, causing haziness. Therefore, wines designated for carboxymethyl cellulose treatment must be protein-stabilized first.

5.13.4.4 Mannoprotein

Mannoprotein, a soluble glycoprotein derived from yeast cell walls, can adsorb onto crystal nuclei, preventing their growth and thus stabilizing tartrates in wine. The effectiveness of mannoprotein as a stabilizer depends on the dosage and type of wine being treated. In white wines, mannoprotein not only stabilizes tartaric acid but also enhances protein stability and improves the wine's organoleptic properties. However, its use in red wines is limited due to interactions with fining agents like saponite, gelatin, and bentonite, which can decrease mannoprotein levels. Consequently, mannoprotein is primarily used to improve tartaric stability in white and rosé wines.

5.13.4.5 Potassium Polyaspartate

Potassium polyaspartate is a macromolecular polymer of the amino acid L-aspartic acid, characterized by active groups such as peptide bonds and carboxyl groups. It is recognized for its chelating, dispersing, and adsorptive properties, making it useful as a scale and corrosion inhibitor, a synergist for fertilizers, and an enhancer in industries like papermaking, printing, dyeing, and laundry. In the winemaking process, potassium polyaspartate plays a vital role in preventing tartar crystal formation by altering the surface properties of the crystals. For high colloidal instability in red wines, pretreatment with bentonite is advised as the final step before bottling.

5.14 Products for Improving Sensory Characteristics

During winemaking, practices and additives legally permitted to improve aroma and flavor in alcoholic fermentation are crucial. Below are some prevalent enhancers and methods used for this purpose.

5.14.1 Additives

Yeast Strains: Various yeast strains, especially non-Saccharomyces, can generate distinctive aromatic compounds. For instance, *Saccharomyces cerevisiae* is renowned for producing esters and higher alcohols that add fruity and floral notes to the aroma.

Nutrient Additions: Nitrogen sources like diammonium phosphate (DAP) and amino acids can influence aromatic compound synthesis. Peptides are also known to improve fermentation kinetics and aroma generation.

Enzymes: Glycosidases and other enzymes can liberate bound aroma compounds from grape precursors, thus enriching the wine's aromatic profile.

Thiols: Certain additives that release or amplify thiols can impart tropical fruit aromas, especially in white wines.

Temperature Control: The temperature of fermentation greatly impacts aromatic compound synthesis. Lower temperatures tend to promote ester production, which results in fruity aromas.

Oxygen Management: Regulating oxygen exposure can modify yeast metabolism and the synthesis of specific aroma compounds.

Lees Aging: Maturing wine on lees (spent yeast cells) can improve mouthfeel and add complexity to the aroma by releasing mannoproteins and other elements.

These additives and methods are commonly employed by winemakers to craft wines with enriched and appealing aromatic profiles [29].

References

1. Tsakiris A (1998) Enology from grape to wine. Psyhalou Publications, Washington
2. Cairns-Fuller V (2004) Dynamics and control of ochratoxigenic strains of Penicillium verrucosum and Aspergillus ochraceus in the stored grain ecosystem. PhD Thesis, Cranfield University, Silsoe

3. Tiralongo J, Wohlschlager T, Tiralongo E, Kiefel M (2009) Inhibition of Aspergillus fumigatus conidia binding to extracellular matrix proteins by sialic acids: a pH effect? Microbiology 155:3100–3109
4. Kassemeyer HH, Berkelmann-Lohnertz B (2009) Fungi of grapes. In: Biology of microorganisms on grapes, in must and in wine. Springer, Berlin, pp 76–78
5. Janex-Favre MC, Parguey-Leduc A, Jailloux F (1993) The ontogeny of pycnidia of Guignardia bidwellii in culture. Mycol Res 97:1333–1339
6. Charoenchai C, Fleet G, Henschke PA (1998) Effects of temperature, pH and sugar con-centration on the growth rates and cell biomass of wine yeasts. Am J Enol Vitic 49:283–288
7. Torelli E et al (2006) Ochratoxin A-producing strains of Penicillium spp. isolated from grapes used for the production of "passito" wines. Int J Food Microbiol 106:307–312
8. Mendoza LM, Fernández de Ullivarri M, Raya RR (2018) Saccharomyces cerevisiae: a key yeast for the wine-making process. A Closer Look at Grapes, Wines and Winemaking. Nova Science Publishers. (pp. 173–201). ISBN: 978-1-53613-289-2
9. Benito Á, Calderón F, Benito S (2018) Schizosaccharomyces pombe biotechnological applications in winemaking. In: Singleton T (ed) Schizosaccharomyces pombe. Methods in molecular biology, vol 1721. Humana Press, New York. https://doi.org/10.1007/978-1-4939-7546-4_19
10. Soufleros E (2012) Oenology, science and technology. Thessaloniki
11. Meena M, Prajapati P, Ravichandran C, Sehrawat R (2021) Natamycin: a natural preservative for food applications—a review. Food Sci Biotechnol 30(12):1481–1496. https://doi.org/10.1007/s10068-021-00981-1
12. Guo L, Li R, Chen W, Dong F, Zheng Y, Li Y (2023) The interaction effects of pesticides with Saccharomyces cerevisiae and their fate during wine-making process. Chemosphere 328:138577. ISSN: 0045-6535. https://doi.org/10.1016/j.chemosphere.2023.138577
13. Pérez F, Ramírez M, Regodón JA (2001) Influence of killer strains of Saccharomyces cerevisiae on wine fermentation. Antonie Van Leeuwenhoek 79(3–4):393–399. https://doi.org/10.1023/a:1012034608908
14. Jimenez-Lorenzo R, Bloem A, Farines V, Sablayrolles J-M, Camarasa C (2021) How to modulate the formation of negative volatile sulfur compounds during wine fermentation? FEMS Yeast Res 21(5):foab038. https://doi.org/10.1093/femsyr/foab038
15. Duncan JD, Setati ME, Divol B (2024) Nicotinic acid availability impacts redox cofactor metabolism in Saccharomyces cerevisiae during alcoholic fermentation. FEMS Yeast Res 24:foae015. https://doi.org/10.1093/femsyr/foae015
16. Labuschagne P, Divol B (2021) Thiamine: a key nutrient for yeasts during wine alcoholic fermentation. Appl Microbiol Biotechnol 105:953–973. https://doi.org/10.1007/s00253-020-11080-2
17. Maicas S (2021) Advances in wine fermentation. Fermentation 7:187. https://doi.org/10.3390/fermentation7030187
18. Lednicer D (2011) Steroid chemistry at a glance. Wiley, Hoboken. ISBN: 978-0-470-66084-3
19. Girardi Piva G, Casalta E, Legras J-L, Tesnière C, Sablayrolles J-M, Ferreira D, Ortiz-Julien A, Galeote V, Mouret J-R (2022) Characterization and role of sterols in Saccharomyces cerevisiae during white wine alcoholic fermentation. Fermentation 8:90. https://doi.org/10.3390/fermentation8020090
20. Aguilera F, Peinado RA, Millán C, Ortega JM, Mauricio JC (2006) Relationship between ethanol tolerance, H^+-ATPase activity and the lipid composition of the plasma membrane in different wine yeast strains. Int J Food Microbiol 110:34–42
21. Walker GM, Walker RSK (2018) Enhancing yeast alcoholic fermentations. Adv Appl Microbiol 105:87–129. https://doi.org/10.1016/bs.aambs.2018.05
22. Fu J, Wang L, Sun J, Ju N, Jin G (2022) Malolactic fermentation: new approaches to old problems. Microorganisms 10:2363. https://doi.org/10.3390/microorganisms10122363
23. Claus H, Mojsov K (2018) Enzymes for wine fermentation: current and perspective applications. Fermentation 4:52. https://doi.org/10.3390/fermentation4030052

24. de Souza TSP, Kawaguti HY (2021) Cellulases, hemicellulases, and pectinases: applications in the food and beverage industry. Food Bioprocess Technol 14:1446–1477. https://doi.org/10.1007/s11947-021-02678-z
25. Moriwaki C, Matioli G, Arévalo-Villena M et al (2015) Accelerate and enhance the release of haze-protective polysaccharides after alcoholic fermentation in winemaking. Eur Food Res Technol 240:499–507. https://doi.org/10.1007/s00217-014-2348-z
26. Maicas S, Mateo JJ (2016) Microbial glycosidases for wine production. Beverages 2:20. https://doi.org/10.3390/beverages2030020
27. Canalejo D, Martínez-Lapuente L, Ayestarán B, Pérez-Magariño S, Doco T, Guadalupe Z (2023) Grape-derived polysaccharide extracts rich in rhamnogalacturonans-II as potential modulators of white wine flavor compounds. Molecules 28:6477. https://doi.org/10.3390/molecules28186477
28. Marangon M, Vincenzi S, Curioni A (2019) Wine fining with plant proteins. Molecules 24:2186. https://doi.org/10.3390/molecules24112186
29. Duc C, Maçna F, Sanchez I, Galeote V, Delpech S, Silvano A, Mouret J-R (2020) Large-scale screening of thiol and fermentative aroma production during wine alcoholic fermentation: exploring the effects of assimilable nitrogen and peptides. Fermentation 6:98. https://doi.org/10.3390/fermentation6040098

Wine Tasting

6

6.1 The Art of Wine Tasting

The basic process for evaluating a wine is what we call wine tasting and you have specific rules and a specific process to ensure objectivity. The process of wine tasting is something that has evolved over the years and changed forms since the multitude of wines available on the market forced consumers and of course professionals to choose the wine that suits them best. Also, the producers choose the wine they wish to market according to the organoleptic characteristics they desire. Consumers in this context have increased their demands and now tasting is one of the essential tools of professionals in order for their product to have a positive response and stand out among the rest of the wines in the increased competition. Now even the very small traditional producers who were once indifferent to perfection have changed and are giving a big base in this sector [1].

The mentality that existed that our wine is the best has been overcome to a large extent, since such a product enters the scale of quality assessment whether we like it or not. A wine that has been created sloppily cannot compete with the rest of the wines that go through strict organoleptic controls and therefore cannot be sold or if it is sold, its price is too low [2, 3]. Consequently, the oenologist's job is no longer limited to maintaining and correcting acidity if he wants the product he curates to stand out and continually win over consumers. Technological progress now makes it possible for a mediocre raw material to be transformed into an excellent wine.

In all of the above, the judge is the consumer taster or the trained taster and his tool is tasting. A tasting can be either simple or complex. In the first case, someone simply tastes the wine to see if he likes it, and in the second case, the tasting process aims at evaluating the quality of the wine [4].

If we want to be objective in the evaluation of a wine, we must follow some rules. The first and most important principle that we all need to understand is that there are some words that we need to delete from our vocabulary when we want to evaluate a wine that we have tasted. The two prohibited words are the word "worse" and the

G. Z. Kyzas, F. Papageorgiou, *Oenology in Practice*,
https://doi.org/10.1007/978-3-031-85531-3_6

word "better." We should not judge a wine based on subjective criteria. It is a fact that each one of us has some preferences when it comes to tastes. For example, a consumer who prefers sweet wines cannot be objective by tasting a dry wine that does not contain sugar. Of course the other way around, generally in a tasting the rules are strict for this very reason, so as not to do injustice to a wine due to personal and therefore subjective preferences [5].

Tasting is an art which also needs a scientific background. Like all other arts, it becomes a combination of innate ability and training. Training should be continuous, persistent, and comprehensive. It is not possible for someone who just attended a seminar to acquire the right to judge wines without further knowledge and without practice. The art of tasting has great power in consumer perception and requires responsibility and maturity in order to practice the profession of taster.

More especially now the taster must be careful and very receptive when trying new flavors. We must adjust our critical view so that it is done in such a way that it does not offend or glorify. Even a wine that we taste and judge to be unsuitable organoleptically, the good taster must be able to understand what is bothering him and what is the reason, and then criticize without exaggeration [6]. Don't always forget that by tasting a wine we judge the work of some people for at least one consecutive year. We have heard the word tasting dozens of times, but what do we mean by this concept?

First we should make it clear that a fixed way of testing greatly helps the objectivity of the judgment. The goal in a tasting is to isolate and enhance the impact of the wine's sensory characteristics—the colors, the aromas, the flavors—in order to distinguish one wine from another.

In essence, think of each new wine we try as a new entry in a database located in our brain. Those who are even more organized can either take notes in a notebook for each wine they taste or even use auxiliary applications (programs) that can be found free of charge and easily installed on the mobile phone or tablet. The same work can be done by taking notes in a notebook that we have exclusively for this purpose. It may sound excessive to take notes on every wine we taste, but it is very useful since, however, we do it there is a maximum limit of flavors that we can hold in our mind, as opposed to an unlimited number of wines that we can taste.

As mentioned before, a consistent way of tasting is the safest way to objectively evaluate a wine. This fixed way of testing includes six stages which are also used by professionals in the field. Of course, we must make it clear that the requirements of a professional taster, who evaluates a wine with a view to its subsequent commercial value and course, are different, and the requirements of a consumer who, in the worst case, not to buy the particular wine again if he does not like it, are different.

At this point it should be emphasized that it is not possible to make a comparison between different types of wines. If, however, there are various categories of wines that need to be appreciated, then the order that is suggested to be tried, in order to achieve the best result, is the following:

1. Dry white wine, light—new
2. Dry white wine

3. Dry rosé wine
4. Red wine—new
5. Dry white wine, aged
6. Dry red wine, aged
7. Semi-sweet white wine
8. Semi-sweet red wine
9. Sweet white wine
10. Sweet red wine

Before describing the steps followed by the taster to evaluate a wine, we should describe a little the preparation of the tasting test in order to get the best results. So initially, the wines we are going to taste should be at the right temperature which is:

1. For white wines 10–14 °C
2. For reds 16–20 °C

6.2 The Tasting Room

The place where the test will take place should have a pleasant temperature of 18–20 °C and in no case greater than 25 °C. The lighting can be either natural or artificial, but it should definitely be sufficient and not excessive. In this way, the taster is able to capture all the nuances of the wine. It is very important that the atmosphere is free from smells (perfumes, flowers, cigarettes, etc.) as well as from noises that negatively affect the judgment of the taster. The tables should either be white by construction or have white tablecloths to create the appropriate contrast. The glasses must be suitable for the wine tasting and meet the minimum characteristics and must not have been washed with detergent because the taste of the wines is affected. Bits of bread, water, and sparkling water should be on the table for the tester to rinse his mouth, but also to create a neutral taste after each test.

The most suitable time to do a tasting is in the morning between 10 and 12. Also, the duration of the test should not exceed an hour and a half and the number of samples should not exceed 15. Finally, for the best results, the tester should be prepared both physically and mentally. If we consider the previous ones as given, we proceed to the next step of the test, which now concerns the wines to be examined. More specifically, before we proceed with the assessment of the content, we should make sure that the wine has not changed organoleptically due to the packaging, and we achieve this by checking the quality of the cork, specifically when the cork crumbles when opening the bottle or is porous and slides with unusual ease in the neck means that we have poor quality cork. Another important problem that cork can create in the wine to be evaluated is that it has altered the organoleptic characteristics of the wine to be evaluated due to negative cork odors due to fungi growing on the cork. The chemical compounds responsible for the unpleasant musty smell are TCA (2,4,6 Trichloroanisole) even at concentrations of 10 ppt.

If the cork comes out with difficulty and is solid without pores, strong cracks, and odors, it means that the cork is of good quality and the wine is.

After we have completed our preparation, we are ready to taste—evaluate the wines by following the following six steps each time [7].

6.3 The Six Basic Steps of a Taste Test

Stage 1 We look at the wine. At this stage, we distinguish the color of the wine (red, rosé, and white). The basic condition of this stage is the good lighting of the space. (It's not very fair to a wine to be in a dimly lit bar trying to judge the hue). After ensuring good lighting, we tilt the glass toward a white surface, for example, a white towel, in order to appreciate what we see. The depth of color, signs that show aging depending on the color of the wine we are looking at, suspensions at the bottom or signs that prove fresh wines and various others (Fig. 6.1).

Stage 2 We swirl the wine in the special tasting glass. With this rotation of the wine in the glass, the aromas become more intense. One reason this is done is because by rotating the contained wine on the walls, the surface area of the wine increases and thus the degree of evaporation of the aromatic compounds increases.

Stage 3 Smell the wine. At this stage we put our nose to the glass and take some deep breaths, at the same time emptying our mind of other thoughts in order to be able to be completely focused on what we are doing. How strong is the scent? What does it remind us of? Fruits, flowers, spices or something else? Does the aging in oak barrels cover the rest of the aromas? We must let our imagination run free.

Fig. 6.1 Wine tasting rose wines

Not a few times we have heard in the description of wine strange words like bubblegum or some food. Indeed, the bubblegum has the aroma of banana and in some food, for example, the cinnamon or pepper that we have added, are aromas that can be detected in some wines [8].

Common Question

The most frequent question asked by consumers is, how can a wine smell of cinnamon, pepper, or banana and forest fruits since it comes from grapes. Very reasonable question with a very simple answer. We smell like a banana the chemical compound that is in greater concentration in this particular fruit and we have named it banana aroma to distinguish it. However, the banana aroma in particular is the chemical compound called ethyl isoamyl ester, and this particular chemical compound can also be detected in wine, not of course in the concentration present in the banana fruit so that the smell is so strong, but in concentrations greater than the so-called determination threshold which is also the lowest limit that a person can understand and smell an aromatic compound. So just as the banana aroma is a chemical compound called ethyl isoamyl ester and by smelling it, we perceive that we smell banana, so the rest of the aromas that we may encounter in a wine are chemical compounds in concentrations greater than the perception threshold. Some very characteristic and common examples of aromatic compounds that we come across are ethanol and methanol which have aromas that are sometimes reminiscent of specific varieties of apples. Another chemical compound called hexanol has the smell of freshly cut grass. Tyrosol has a honey aroma, etc.

Stage 4 Take a sip of wine. At this stage we will finally taste our wine, having seen it, stirred it, and smelled it, so we take a slightly larger than normal sip. We do not immediately swallow the sip we drink, but hold it in our oral cavity for about 5 s, so that the wine covers every surface of the mouth (tongue, palate, and cheeks).

Stage 5 We swirl the wine in our oral cavity. At this stage, the taster swirls the wine in his mouth like we do, for example, with mouthwash. This is done to increase the sense of taste, smell, and texture in the mouth. As with stirring in the glass, this type of stirring in the oral cavity increases the contact surface and the sensations become more intense. Another additional reason why this stage is very important for the overall assessment is that during rotation the body temperature heats the wine, the rate of evaporation of its aromatic components increases and thus its aromas are concentrated in the olfactory nerves.

Stage 6 We taste the wine. In this last stage, we swallow the wine and for this we use the word taste. When the wine is swallowed, what we call the aftertaste does not immediately disappear. The aftertaste is essentially the taste impression that remains after the wine has left the mouth. We don't want the aftertaste of the wine to be so short that we forget about it, but neither do we want it to last long and persistently. At this stage, it is also criticized.

We Either Spit or We Don't Spit the Wine We Taste
A key difference that exists when tasting a wine between an amateur and a professional taster is that the professional must spit out the wine they are tasting, while the amateur consumer does not have to spit it out. This is because the professional tasters who in a tasting test are obliged to taste a large number of wines, it is not possible to drink all the wines they taste since apart from getting drunk and not being able to continue their work, their final result is also affected crisis due to the accumulation of alcohol in the body. However, consumers who want to enjoy the wines they taste are not obliged to spit out the wine, since in principle the number of wines they will taste is not that great and then the purpose of a tasting of a group of wine friends is mainly entertainment.

6.4 The Five Senses

For the overall appreciation of the wine, we rely on four of the five total senses we have. Sight, taste, smell, and touch of the oral cavity help us appreciate the differences and similarities that one wine has from another. The only sense out of the five in total that does not play a role in wine tasting is hearing. Perhaps the only case where hearing also participates in the taste impression of a wine is when we taste sparkling wines or champagnes. But let's see in detail what each of the above senses helps us with [9].

6.4.1 Sight

It is perhaps unnecessary to describe what we mean by the sense of sight since even now someone reading this text is using his sight. For formal reasons, however, we should mention that vision means visual perception. The organ of perception is the eyes, while the object of perception is light. Sight is considered the most important of the other senses, because with it the external space is immediately perceived. About 30% of the human brain is involved in the processing and interpretation of visual stimuli. In a taste test, with the help of sight, we recognize the color of the wine, the depth of the color, the clarity, the brilliance, and the fluidity of a wine and we can classify the wine in terms of:

Color—white, pink, red
Color depth—light, medium, dark
Clarity—cloudy, clear
Luster—dull, shiny
Fluidity—liquid, viscous

All three basic shades of wines consist of a large number of shades. At this point, we should emphasize that the senses work simultaneously and complementarily, to render reality. Sight as well as hearing allow a better perception of extracorporeal space, especially when it comes to distant objects. Also, sight allows direct supervision of the stimuli of the other senses. For example, if we smell something with a bad smell, we can look at it and see if it is moldy.

In a tasting, the senses are difficult to isolate, but the professional taster has a duty to isolate these senses since during the evaluation there is usually a grading and it is unfair to affect the overall score of a wine because one of the senses gave us the information that it is for an inferior wine. To be more specific, a wine that visually gives us the information that it has sediment due to tartar salts does not mean that it is a bad wine in general. There are endless examples of wines that visually may predispose a strict judge to it but it is an excellent wine.

6.4.2 Taste

Taste is one of our basic senses, the instrument of perception is the tongue, while the object of perception is the chemical composition of food and drink. On the tongue, but also inside the cheeks and lips, there are many cells, the taste buds. Each taste bud is specialized for only one of the four tastes. In general, calyxes of all four types are present everywhere on the tongue. However, in the periphery of the tongue, there are five regions, each with a high concentration of calyces of a particular taste. On the tip of the tongue is the sweet. The remaining areas appear symmetrically on the left and right of the tongue. From the tip of the tongue to the throat, the remaining areas are salty, sour, and bitter. With taste we recognize the sweetness and acidity of a wine in the categories we see but they are not the official categories that exist according to the legislation. They are the categories that a person can perceive based solely on their taste and not on chemical analyses.

Sweetness—dry, slightly sweet, sweet
Acidity—medium acidity, bitter, acidic
Salty—it is extremely rare to encounter salty taste in a wine
Bitter—we rarely find the bitter taste in a wine that has not undergone serious alterations.

6.4.3 Smell

Olfaction is the sense of identifying substances through the volatile molecules they emit and come into contact with. The identification/recognition of the various substances is done by special nerve endings inside the nose, which send the olfactory message to the brain. The sense of smell can provide information about the chemical composition of the environment. The organism based on its experience can identify what it is that smells. Substances are identified by special sensors that are nerve endings. During breathing or in general air flow particles from the substance are captured by the nerve sensors and they identify the shape and size of the particle sending the information to the brain. Each sensor can identify up to a hundred different particles. In general, the larger the surface area with the sensors, the more sensitive the sense of smell. In this way, and in a wine tasting, these sensors lead us to various conclusions. The most common characteristics we describe when smelling are the intensity of fruits, flowers, spices, and others as well as the aroma of the barrel during aging.

The specific sense to get to the point of evaluating wines that are graded must be trained a lot. We can all smell something, very few can identify what it is they smell. This applies even to well-known aromas such as banana or strawberry. It is very important that professional tasters regularly train with different aroma kits so that they can continuously enrich their olfactory images and be able to distinguish aromas in very small concentrations. It is a painstaking process and requires excellent training to reach the point of being called a professional taster. No matter how many seminars we attend if we do not practice regularly it is almost certain that we will remain stagnant and be content with recognizing a small number of scents out of the thousands that exist in nature.

By smell we classify wines into two categories:

Intensity of characteristic aromas—mild, aromatic, intense
Barrel Wood Aroma—no beech, light beech, strong beech

6.4.4 Touch (Oral Cavity)

With the touch of the mouth, we understand some of the most important characteristics of the wine such as body, tannins (Only in red wines), and carbonation:

Body—light, medium, heavy
Tannins—silky, velvety, rough
Carbonation—non-carbonated, lightly carbonated, carbonated

We never compare different types of wines.

Understandably, white and red wines have different characteristics and rosé wines have different characteristics. It is not correct to compare with the same criteria different types of wines. The same applies to wines with different sugar contents. It is not fair to judge a dry, sugar-free wine with the same criteria as a semi-dry, semi-sweet or sweet wine. If we want to be fair and correct, we should always compare wines belonging to the same category, for example, white dry or red dry, etc. [9, 10].

Many times the question is asked about the most important feeling of the ones we mentioned. It is logical that every sense has its importance. But if we had to single out one of them, I would choose the sense of smell. But the sense of smell is the one that gives us the most information about the aromas that can emerge from a wine. To understand exactly the role of smell we can do what we call the "nose test."

The Nose Test

Basically this test is done to understand the difference between taste and smell.

1. Close the nose and take a sip, for example, fresh orange juice. We continue to keep the nose closed for several seconds after swallowing the orange juice.
2. What we understand from this test is that when we swallow the juice and have the nasal passage closed, we only perceive what the tongue tastes. In the case of orange juice sour and sweet depending on the ripeness of the oranges we had.
3. When we leave our nose and free the nasal passage, the air starts and passes freely from our palate to the olfactory nerves of the nasal passage and we directly understand the taste of orange juice.

6.5 Types of Taste Tests

There are various types of taste tests that can be carried out, the basic condition of which of these types we choose is the parameter we wish to focus on. For example, if we judge that the taster should not know the wines he will taste, then the blind test is appropriate and not the obvious one. Another parameter that affects the character of a tasting is the purpose of the tasting, based on this parameter the tasting is classified as professional or amateur. Otherwise the number of tasters determines whether the tasting is individual or group. In other cases, the determining factor is the type of wines to be tasted, based on this parameter a tasting can be characterized as either horizontal or vertical. But the last two categories are not the only ones based on the parameter the type of wines we taste, the type of wines we taste can be white, red, rosé, dry, sweet, fresh, aged, single varietal, etc. Another parameter that determines the type of tasting is the panel tastings that judge whether a wine meets the conditions to be designated PDO. Another case of taste testing is when we try samples consisting of different percentages of the same wines in order to arrive at a blend that will satisfy us to market. Another type of tasting is tasting a wine to judge whether it goes well with a range of foods and creating guide categories that will make it easier for professionals and consumers to choose their wine according to the type of food. But let's focus on the basic categories of dividing a tasting since as one understands as long as there is human imagination one cannot put strict limits on the type of categorization of a tasting test [9–11].

6.5.1 Horizontal and Vertical Wine Taste

In the case of horizontal tasting, the parameter we focus on is the year of production of the wine. More specifically in a horizontal test same wines (variety) of the same year. The purpose of this test is the qualitative distinction between wines created in the same year. If the number of samples is satisfactory, then conclusions can be drawn for the total production of the year under consideration [11].

In the case of a vertical test, the parameter we focus on is the year of production of the wine. More specifically, in a vertical test, we taste the same wines (same variety) that have been vinified in different years. The way these tests are carried out have a strict methodology in order to reach safe conclusions. More specifically, we always start from the most recent year and go backward in time. With the vertical tests, we can understand the evolution of a wine or a specific variety, we can identify the maximum duration that a wine can remain in a bottle before judging it to have negative organoleptic characteristics or finally identify a year which we can characterize either as better or as worse for the specific wine.

6.5.2 Open and Blind Wine Taste

In an overt tasting the taster knows the identity of the wine he is going to taste. In a blind test, the taster either knows absolutely nothing about the wine he is going to taste or knows very few characteristics in order to reach safe and fair conclusions (Harvest, winemaking technique, variety, etc.)

6.5.3 Individual and Group Wine Tasting

The distinction between an individual and a group tasting does not need further analysis since it is clear that the number of tasters is the determining factor of characterization. But what needs a lot of analysis is whether the conclusions of an individual test are safe since in this case it is the taste and judgment of an individual that leads to conclusions which in many cases are published and influence consumer preferences appropriately or inappropriately the range of the tester.

In the group test, the participants are definitely more than one and in most cases they are divided into groups headed by the most experienced tester. Group tastings are prevalent in wine competitions as well as various wine rating media. These groups are homogenous and consist of various professional categories (Oenologists, winemakers, wine journalists, etc.). Group tastings are certainly more subjective and lead to more secure conclusions without this meaning that we nullify the value of individual tastings.

6.5.4 Professional or Amateur Wine Tasting

The defining factor of qualification in these taste test categories is adherence to strict regulations and methodology. It is clear that when a tasting is done among wine-loving friends, the rules are often not met and the conclusions are not safe to publish in a journal since maximizing satisfaction is the purpose of such a tasting. But in a professional tasting, the rules and methodology must be strictly followed since the conclusions directly affect businesses and producers, positively or negatively.

6.5.5 Wine Tasting During Production

In these cases of tasting, the oenologist at all stages of the winemaking evaluates organoleptically the evolution of the must into wine at regular intervals in order to intervene in case he deems that some intervention is needed. Of course, the oenologist's basic tool in this type of tasting is the chemical analyzes that focus the oenologist's attention on specific negative or positive organoleptic characteristics.

6.6 Flavors and Aromas

We must know that there are thousands of things that we can smell but only six that we can taste [12]:

1. *Sweet*—honey
2. *Salty*—salted water
3. *Sour*—lemon
4. *Bitter*—tea
5. *Fat*—butter
6. *Umami*—monosodium glutamate

The first four flavors have been known for centuries, the last two are more unknown and recently discovered. Fat is a recently discovered near-perceptible taste associated with fats in foods. Umami is basically the good taste or the deliciousness that is triggered by specific chemicals (glutamates and amino acids) and was discovered by Japanese scientists in their quest to find out why algae taste so good.

6.6.1 Sweetness and Acidity

In wine tasting, the flavors we look for and appreciate are sweetness and acidity. We are not looking for other flavors since the wine contains neither salt, nor fat, nor bitter ingredients. Many wines contain umami, but this is not easily perceived and we will not deal with it. We can rarely encounter a suspicion of saltiness in wines that come from vineyards located near the sea or used for irrigation with water with a high salt content [13].

6.6.2 Sweetness

In the world of wine, sweetness is measured in grams of sugar per liter. The following examples give us an idea of where different types of wine stand compared to other drinks. In essence, we compare the sweetness of characteristic wines with drinks well known to all of us.

- Dry Cabernet Sauvignon: 2 g/L
- Chamomile with a spoonful of sugar: 12 g/L
- Slightly sweet Merlot: 15 g/L
- Orange juice: 85 g/L
- Sweet Porto wine: 100 g/L
- Ripe grape juice: 225 g/L

Sweetness is perceived as a sweet sensation in contact with the tongue mainly at its tip where the taste receptors are densest. If we want to be a little more detailed about sweetness, we should mention the fact that the main sources of sweetness in wines are glucose and fructose. Fructose is sweeter than glucose. The sensation of sweetness can be created by ethanol as well as glycerol.

Most wines out there have no perceptible sweetness and in oenological terms we describe them as dry. Many times there is a confusion on this subject since the word dry that we use to express the almost zero sugar content in wines, in everyday life means a lack of moisture and not a lack of sweetness. So we have to keep in mind that when we say a wine is dry it has a very low content of unfermented sugars. But let's see what are the categories of wines we have according to the legislation in terms of sweetness:

- Dry—up to 4 g/L unfermented sugars
- Semi-dry—from 4 to 15 g/L unfermented sugars
- Semi-sweet—from 15 to 45 g/L unfermented sugars
- Sweet—more than 45 g/L unfermented sugars

More generally, however, the majority of wines around the world are dry, since in principle they are easier to produce, have a longer shelf life and combine more pleasantly with food.

6.6.3 Acidity

As we mentioned before in the wine tasting, the flavors we look for and appreciate are sweetness and acidity. The question of acidity in wines is much more complex than that of sweetness. The acidity is perceived on contact as a sour sensation, almost immediately causing salivation. The main acids that affect the acidity in a wine are tartaric acid, citric acid, malic acid, and lactic acid.

The sensation of acidity of a wine depends on many factors. For the sake of convenience, I will consider pH as the only influencing factor for the acidity we feel and thus I will give examples of drinks with a pH greater or less than wine so that we can understand the levels at which it varies. But before we give the examples, I should emphasize that the pH scale measures acidity in a different way than usual, i.e., the higher the number, the less acidic it is.

But let's see where wine stands in relation to other drinks and beverages:

- Lemon juice pH: 2.4
- Vinegar pH: 2.9
- Wine pH: 3.0–4.0 depending on the wine
- Beer pH: 4.5
- Milk pH: 6.5
- Seawater pH: 8.0
- Soap pH: 9.0–10.0

6.6.4 Stew and Bitterness

The feeling of stiffness is a sensation that is quite annoying when it is intense. When the taster tastes a wine that is strongly astringent then a dry mouth is produced accompanied by a sensation of fine sand in the mouth. It is therefore understandable that they do not want a wine to be strongly astringent because it does not give us satisfaction when drinking it. The astringent taste is usually found in red wines that are rich in tannins. The tannins come from the skins and skins of the grapes, so in red winemaking where we have a long contact time of the grapes with the must we also have an enrichment of phenolic compounds. Tannins can also come from the wooden barrels that the wines are aged in. Finally, another source of tannins is their use during the production process by oenologists to improve the quality of wines.

The phenolic compounds responsible for the acrid sensation are also responsible for the bitter taste. So during the production we have to be very careful and check our production often in order to create wines that are pleasantly consumed and are not bitter and astringent to an annoying degree. In any case, it is very important to check the taste balance by trained oenologists so that there is a balance in the flavors and sensations that a wine creates [14].

6.6.5 The Smell of Wine

More generally, we could say that there are many chemical compounds that, when they are in a concentration greater than the so-called perception threshold, bring to mind various fruits, flowers, animal smells and beeches, sometimes more and sometimes less. More than 600 chemical compounds are responsible for the various aromas we perceive in wines.

White Varieties

- Malagouzia—we smell rose petals and citrus fruits
- Assyrtiko—we recognize lemon blossoms and orange peels
- Vilana—green apple
- Muscat—herbs, rose, spearmint, mint
- Chardonnay—grapefruit and exotic fruits
- Sauvignon Blanc—melon and peach

Red Varieties

- Agiorgitiko—cherries, plums
- Xinomavro—thick tomato paste, olive
- Black laurel—raisins and figs
- Mantelaria—fruit and compote
- Cabernet Sauvignon—fruits of the forest

Novice wine tasters should start by assessing the overall intensity of the aromas (fruit, flowers, etc.) and not try to assess specific fruits, flowers or aromas, because due to the complexity of the aromas, there is a high chance that they will end up with wrong assessments and besides, there is no reason to make their lives difficult. Experience at this point is a valuable ally of the connoisseur. Due to the complexity of the aromas found in a wine, it is difficult to describe a wine in one sentence as is done with most foods. So the taster has a difficult task as he has to isolate among hundreds of perfumes one or more perfumes that stand out, identify them and describe them effectively. The taster simply tries to identify the type and intensity of the most characteristic aromas of a wine. Nine categories of aromas have been created that one can detect in a wine. So, once the tester identifies the general category, he then tries to identify which fragrance it is [15].

These categories are as follows:

1. Floral scents
2. Aromas of fresh fruit
3. Aromas of dried fruits and nuts
4. Aromas of dry grasses and foliage
5. Aromas of roasting and smoke
6. Aromas of herbs and spices
7. Balsamic aromas

8. Animal scents
9. Aromas of various foods

6.6.6 Barrel Aroma

A very characteristic aroma present in all aging wines is that of barrel. The barrel chapter is a separate chapter in the literature of aromas and there is no reason to delve more simply to keep the general rule that the longer a wine has been aged in a barrel, the more intense the aromas of beech or aging aromas are usually. Of course, in addition to the aroma of barrel, a wine that matures in a barrel can acquire aromas of vanilla, spices.

6.6.7 Negative Wine Odors

A wine as we analyzed before can have all these positive aromas that will attract us to prefer it, but there can also be unpleasant odors that reveal serious alterations. These alterations can either come from the entire production or exclusively in one or a small number of bottles of the entire production. This should be mentioned since the responsible tester before coming to safe conclusions must make sure that it is the whole production or not. This can be done in a very simple way. In the event that unpleasant odors are detected in one wine, we can open several bottles of the same wine, if the unpleasant odors are also detected in the other bottles, then the problem is most likely found in the whole of the particular lot of the particular label. Otherwise it is a statistically justifiable one-bottle error. However, for so long we refer to negative odors without first clarifying what are considered negative odors for a wine.

6.6.8 Oxidation

As oxidation we perceive the smell that acetaldehyde has and always in such wines it has been affected in addition to the taste and color at an advanced stage and the highlights during the reflection of light at an early stage. Oxidation faults are found in bottled wines where either the storage conditions were problematic or the materials used during bottling were not suitable (Cork, bottle, capsules).

6.6.9 Acetone Odors

The smell reminiscent of acetone used to bleach nails in a wine is due to the presence of ethyl acetate in concentrations greater than 150 mg/L. The main source of this odor is microbial spoilage.

6.6.10 Acetic Acid

When we get to the point of detecting the annoying smell of acetic acid organoleptically in a wine, then we have to take it for granted that this wine has already reached concentrations of acetic acid greater than 0.8 g/L and unfortunately the satisfaction that the consumer receives by drinking a such wine is almost zero.

6.6.11 Grassy Smells

The smell of freshly cut grass may not be so unpleasant, but it is a smell we do not want to find in our wine. This smell is due to aldehydes and alcohols that have six carbon atoms.

6.6.12 Sulfite

The smell of sulfite is a smell which is due to the addition of sulfur dioxide intended for the preservation of wine. The oenologist must be very consistent in the addition of sulfite since we must in principle ensure the preservation of the wine but at the same time the concentration of free sulfite must not exceed the perception threshold. The smell of sulfite in recently sulfited wines cannot be avoided, but it is temporary and disappears after a few days.

6.6.13 Smell of Butter

And in this case the smell of butter one cannot say with certainty that it is a negative smell, but it is an undesirable smell for wine. The particular smell is due to diacetyl which is a by-product of the metabolic activity of yeasts and some bacteria.

6.6.14 Reducing Odors

By the concept of reduction is meant a situation in which the wine is closed in a container and does not exchange gases with the environment. Reductive odor is a very negative odor and we can present it with the smell of cage egg or manure. This smell is due to hydrogen sulfide and various other sulfur compounds that can be created at any stage of production.

6.6.15 Smell of Geranium

The smell of geranium comes from the sorbic acid that is added to wines to preserve them and avoid fermentation. However, in addition to the addition of sorbic acid, it can also be produced by certain lactic acid bacteria. In the second case, the concentration of the sorbic acid produced is hardly greater than the perception threshold, so it is almost impossible to detect it organoleptically due to the lactic acid bacteria.

6.6.16 Fuel Odors

Fuel odors are due to the higher alcohols produced during alcoholic fermentation. To be noticed they should have a concentration greater than 250 mg/L. It is not that common an odor and only well trained oenologists/tasters can detect such odors. Under no circumstances do we want a wine that gives us the feeling that we are about to drink a wine that resembles fuel.

6.6.17 Cork Smell

The cork smell is clearly from the cork we use to cap the bottles. If a wine has a cork smell, it is immediately noticeable when it is opened. This smell is similar to the smell of a mop that has been wet for days. This can come from a large number of chemical compounds, but the most common is 2,4,6-trichloroanisole or otherwise TCA. The reason we may have the production of this chemical is microbial infestations of the cork. Such odors are very common and statistical studies have proven that wines using 5% natural cork have problems of such odors.

6.6.18 Burning Smell

Such odors can appear in wines that have been exposed to very high temperatures either during their storage or during their transport.

6.6.19 Mouse Smells

Those who hear the smell of a mouse for the first time will almost certainly wonder how it is possible for someone to know what a mouse smells like. It's a fair question, but researchers have found that tetrahydropyridine odors created by several species of fungi are found on the skin of mice and produce a specific odor. This odor can also be detected in a wine that has been affected by fungi mainly Lactobacillus and Brettanomyces.

6.6.20 Smell of Naphthalene

The smell of naphthalene is a characteristic aroma and can be detected in a wine. This smell has roots that come from the cultivation of the grape and not from the production process of the wine.

6.7 Wine Tasting Glass

There are various glasses one can use in a tasting which can vary in shape and quality. In any case, the glass should be thin and completely transparent so as not to affect the visual inspection of the wine. The stem of the glass should be of a size that allows easy grasping but our hand does not hide the contained wine. The foot of the glass also has the role of a lever for creating circular stirring, a movement that is mandatory in order to have a correct aromatic image of the wine we are tasting [16] (Fig. 6.2).

Also, the fact that the glass is held by the leg prevents the transfer of heat from our hand to the wine, which would significantly affect the results of the taste test. It should have a capacity of 150–300 mL and the mouth should be closed to create an obstacle to the escape of aromatic components from the glass. A usual amount of wine we put in the glass is about 180–200 mL.

Red and rosé wines are usually served in larger glasses than white wines because they have stronger aromas. In the event that a white wine, which has milder aromas, is incorrectly served in a large glass, there is a good chance that we will consider it

Fig. 6.2 Wine glasses

Fig. 6.3 Wine decanters

aromatically weak. Sparkling wines are served in thin and thin glasses because they lose their carbonation faster on a large surface. In cases of sweet wines as well as fortified wines, half portions are served and usually in smaller glasses.

6.7.1 Pouring the Wines into Decanters

We've all seen in a restaurant or at a tasting where the contents of the wine are poured into decanters like the ones in the photo just below (Fig. 6.3).

This process is common in aged wines and is done for three reasons:

- To remove the sediments created during aging from the wine.
- To contact the wine with oxygen and stir it during emptying to remove possible reduction aromas.
- To breathe wines that are young and contact with air helps the flavor to develop [17].

References

1. Galizzi MM (2013) Wine judging and tasting. In: Giraud-Héraud E, Pichery MC (eds) Wine economics. Applied econometrics association series. Palgrave Macmillan, London. https://doi.org/10.1057/9781137289520_7

2. Caggiano G, Galizzi MM, Leonida L (2012), Who is the expert? An analysis of awards to Italian wines. Discussion paper. London School of Economics
3. Combris P, Lange C, Issanchou S (2006) Assessing the effect of information on the reservation price for champagne: what are consumers actually paying for? J Wine Econ 1(1):75–88
4. Galizzi MM, Buonanno P, Caggiano G, Leonida L (2008) Expert and peer pressure in food and wine tasting: evidence from a pilot experiment. Enometrica 1:51–68
5. Blanck JL, Cogan-Marie L, Agnoli L (2019) Importance of tasting room activities and staff training in emerging wine regions: the case of Northern Virginia. In: Sigala M, Robinson RNS (eds) Wine tourism destination management and marketing. Palgrave Macmillan, Cham, pp 497–513. https://doi.org/10.1007/978-3-030-00437-8_31
6. Carmichael BA (2005) Understanding the wine tourism experience for winery visitors in the Niagara region, Ontario, Canada. Tour Geogr 7(2):185–204
7. Marlowe B, Brown E, Zheng T (2016) Winery tasting-room employee training: putting wine first in Oregon. J Qual Assur Hosp Tour 17(2):91–102
8. Pozo-Bayón MÁ, González CM (2024) Wine analysis and testing techniques. Springer, New York. https://doi.org/10.1007/978-1-0716-3650-3
9. Gere A, Kókai Z (2024) Wine descriptive sensory profiling. In: Pozo-Bayón MÁ, Muñoz González C (eds) Wine analysis and testing techniques. Methods and protocols in food science. Springer, New York. https://doi.org/10.1007/978-1-0716-3650-3_11
10. Criado C, Muñoz González C, Pozo-Bayón MÁ (2024) Time-intensity methodology for wine flavor evaluation. In: Pozo-Bayón MÁ, Muñoz González C (eds) Wine analysis and testing techniques. Methods and protocols in food science. Springer, New York. https://doi.org/10.1007/978-1-0716-3650-3_14
11. Perez-Jimenez M, Munoz-Gonzalez C, Chaya C, Fernandez-Ruiz V, Alvarez MD, Herranz B, Pozo-Bayon MA (2022) Insights on the effect of age and gender on in-mouth volatile release during wine tasting. Food Res Int 155:111100. https://doi.org/10.1016/j.foodres.2022.111100
12. Kemp SE, Ng M, Hollowood T, Hort J (2018) Introduction to descriptive analysis. In: Descriptive analysis in sensory evaluation. Wiley, New York, pp 1–39. https://doi.org/10.1002/9781118991657.ch1
13. ISO (2012) ISO 8586:2012 sensory analysis—general guidelines for the selection, training and monitoring of selected assessors and expert sensory assessors. ISO, Geneva, pp 1–38. https://www.iso.org/standard/76667.html
14. ISO (2011) ISO 3972:2011 sensory analysis—methodology—method of investigating sensitivity of taste. ISO, Geneva, pp 1–10. https://www.iso.org/standard/50110.html
15. ISO (1994) ISO 11035:1994 sensory analysis—identification and selection of descriptors for establishing a sensory profile by a multidimensional approach. ISO, Geneva. https://www.iso.org/standard/19015.html
16. Heymann H, King ES, Hopfer H (2014) Classical descriptive analysis. In: Novel techniques in sensory characterization and consumer profiling. CRC Press, Boca Raton, pp 18–26
17. Parr WV, Schlich P, Theobald JC, Harsch MJ (2013) Association of selected viniviticultural factors with sensory and chemical characteristics of New Zealand Sauvignon blanc wines. Food Res Int 53(1):464–475. https://doi.org/10.1016/j.foodres.2013.05.028

MIX
Papier aus verantwortungsvollen Quellen
Paper from responsible sources
FSC® C105338

If you have any concerns about our products,
you can contact us on
ProductSafety@springernature.com

In case Publisher is established outside the EU,
the EU authorized representative is:
Springer Nature Customer Service Center GmbH
Europaplatz 3, 69115 Heidelberg, Germany

Printed by Libri Plureos GmbH
in Hamburg, Germany